RIDING
Through Life

with
Bud Colbow

[signature]

by Jesse Ziegler

Canadian Cataloguing in Publication Data

Ziegler, Jesse, 1970-
 Riding Through Life with Bud Colbow

 ISBN 0-919537-48-0

 1. Colbow, Bud, 1925- 2. Farmers – Saskatchewan –
Biography. 3. Farmers – British Columbia – Vancouver
Island – Biography. I. Title.
S417.C65Z53 1999   630'.92   C99-910509-4

Designed and printed by Kask Graphics
Campbell River, BC, Canada

In Memory of Dwight

A few words on Bud...

"I've been fortunate to have known Bud as a good friend for 40 years. I have always admired and valued his wit, and his down to earth philosophy on life."

Darwin Kvisle

"I have been a part of Bud's life ever since I was a kid in Saskatchewan. From the days of playing in the back of his truck, out at Grandpa's farm, to stealing pickles out of his basement in B.C. (Boy, was Phyllis mad!) I'd like to say that, along with being my favorite uncle, Bud has been one hell of a friend to me and my sisters over the years. Thanks, Bud."

Bill Paisley

"I met Bud's father first in 1946. I later met Bud at Henry Schultz's auction sale, and over the years he has helped my family out in many ways.

Bud has handled thousands of cattle out of the Valley, hauling them for friends; people in need. He'd do anything for anyone. You just don't find people like Bud anymore. You couldn't wish for a better neighbour, or a better friend than Bud Colbow."

The Gunter Family

"My wife and I have known Bud for over thirty years. He's always good for a quick retort. He enjoys life, loves to make a deal and is always ready to pass on good, practical advice."

Dave & Kaye Hansen

"I well remember meeting Bud Colbow at the Livestock auction in Courtenay. The day was clear, but the wind still had a bit of winter bite and the morning ice was not long gone from the potholes in the bleak gravel auction yard.

I had spent everything I had and more, on the set-up of my farm vet practice and was pretty anxious about whether or not I was going to make a go of it.

Bud had his back to me. He was wearing a tractor cap and jeans, with his jacket buttoned up against the cold; his hands thrust into his pants pockets. He didn't turn around until I had climbed out of my van and said "hi".

I knew that he was an old-timer in the area and I asked him for his thoughts on my outlook.

He kind of measured me up for a moment and said, "Work hard, do a good job and you'll be so busy you won't know what to do."

Turned out to be true."

Dr. Pat O'Brien, DVM

Bud grew up in a generation that had to learn the harsh realities of life at a very young age. It's hard for someone that's not from that generation to put the severity of the time in perspective. Bud learned these lessons very young and very well. He learned that good sound business decisions could be made with your word and a firm handshake. To this day when Bud shakes your hand, it's like getting grabbed with a pair of vise grip pliers and his word is as good as if it was written in stone.

The best way I can describe Bud, is to quote a line from one of the poems I have written about him.

"Bud believes that hard work and long days
never hurt no one, and it's hard to argue
when you've seen all that he's done."

I'm very proud to be able to call this man my uncle. Thanks for everything Uncle Bud.

Drew Pederson

Forward

"Bud"

W.L. Colbow is his name,
buying and selling livestock is how he got his fame.
He started as a young boy in Saskatchewan,
in a one horse town called Nipawin.

He moved to the west coast, as a young man,
and tried logging for a while, but it wasn't his hand.
He had to get back into livestock haulin',
because old Bud just loves when them old cows are bawlin'.

He's hauled cows and horses from all over BC,
and brought them back to the auction known as CCC.
Bud can guess the weight of an animal before it gets
on the scale,
the only thing that will throw him off is a pile of shit
stuck on one's tail.

Just about every critter that comes to the auction,
Bud's got a story to tell,
and it's usually what helps that animal to sell.
If he doesn't know a story about a certain cow,
he'll pull one out of his head and have it ready right now.

Bud's been around for quite a few years,
hell that's an understatement, he's been around
as long as Simpson and Sears.
His mind's like a computer, he works numbers so fast,
he makes it so easy like it's not even a task.

Some of the younger generations, that are such duds,
they should take lessons from someone like Bud.
He believes hard work and long days never hurt no one,
and it's hard to argue when you've seen all that he's done.

He's about the best livestock man you'll see,
I just hope a little of Uncle Bud rubs off on me.

By Drew Pederson
December 22, 1997

Contents

Author's Note

I first met Bud in November 1997 interviewing him for a monthly feature, for *In Focus* Magazine. Ironically, I stated in the article that Bud's life experiences would be sufficient to fill a book. Months later Bud proposed just that. The result is 'Riding Through Life'.

Writing this book has been challenging, but rewarding. Knowing Bud has been a real experience, and I feel I only began to know him truly when the book was near completion.

'Riding Through Life' is Bud's story in his own words, written the way he told it. To anyone not mentioned along the way, we apologize.

I am so grateful to my family and close friends for their encouragement and support, especially Mom and Clark for pushing me to finish it. Thank you Mumsy, for your guidance and endless hours of editing! A big thanks to Kask Graphics for all of your advice.

Thank you, Bud, for sharing your story.

1

In the Beginning

I don't remember a lot of the depression. I was young. But I don't think it hit as hard in B.C. as it did in the prairies. Here, all you needed was a little bit of ambition and creativity and you could find food or make work for yourself, hunting or berry picking. You'd freeze to death during those prairie winters if you didn't have a roof over your head.

In Brooking, Saskatchewan, where I was born, it was a common sight to see hobos passing through from the east, and often I'd find one sleeping up in our hayloft. They'd work on the roads to pay their taxes, which amounted to about ten dollars a year. We always had one staying at our house because even when there wasn't much work to do the hobos were happy just to have board and a breakfast. My two older sisters, Mary and Anne, and I attended Brooking School, which taught about twenty-five kids. I had a couple of friends there and we'd hang out a bit, but I was never much of a kid for playing at any time in my life. I did very little playing.

Dad was a blacksmith, and with the economy being the way it was customers could no longer pay him with money. Instead, they would trade things. At one point we wound up owning eleven horses. I can remember people having to put their vehicles up on wooden blocks to prevent the tires from rotting and they'd ride their horses to town. The price of petrol must have been outrageous.

When my seventh birthday rolled around, we had to leave Brooking. Dad was an honest, hard working guy, and he didn't realize that come harvest time all the bankers and their dogs would be in there to scoop up our crops. I guess my dad knew that if we remained in Brooking there'd be no getting ahead. That hot summer we joined the other trains of people who were heading north to Nipawin. Along with their belongings, they carried hopes of work, and hopes of a better life. My family had three teams of horses with hay racks, a couple of cows, and my very first horse, a brown bay horse named "Doll", which I'd received when I was six.

Mom took us kids in the family car, a 1928 Model A, which we knew we were lucky to have. Dad had purchased the car when times were better, before the crash came in '29. That old two-door was packed full with all of our belongings. My sisters and I felt safe travelling with Mom. She was quite a woman. With her fiery red hair and her little frame, she appeared small, but she was tough. I remember she would put a clove of garlic in her boot, and said that her body would absorb it. By the smell of her breath, we believed her.

The days were long, the distances traveled were short, and the journey seemed to take forever. On a slow day we'd cover eight or nine miles, but if we came across a spot where there was water we'd have to stay there for a couple days. At least until the horses were rested up. The next day we'd make up for lost time by going twenty-five or thirty miles. The cows usually didn't fare so well. Dad had fixed the horses up with iron shoes, but had nailed only leather to the cows' hooves, and it didn't last long.

We had our trusty dog, Spot, by our side. He lived on crows on that journey and his coat was as black as coal so by the time we got to the valley we renamed him "Blackie". The horses had to fend for themselves, usually managing on Russian thistles.

Nowadays, a journey like that would probably only have taken about five hours. It took us eight weeks. It was the middle of July, and at night Mary, Anne, Mother and I would sleep under the stars. Dad was a week or ten days behind us and we didn't see him again until we arrived at the quarter section of land which he'd made a deal on, up in Nipawin. We ate whatever we could find, which wasn't much. With nothing to catch, things were pretty scarce. But the weather was warm and if it rained we had the car to sleep under. It was tough going and Dad was always concerned about the horses. He really looked after his stock so he had to stop whenever there was a little bit of food for them. At one point we met up with a farmer who needed a couple of cows to work his land. He offered my father seventeen dollars for two cows, and said, "God knows when I'll ever be able to pay you, but I promise that somehow I will." The deal was made with a handshake, and we moved on. Five years later, the old guy paid my dad back. The handshake meant everything when I was a boy, and for me, to this day, it still does.

2

Life in Nipawin

I liked Nipawin where I went to Sunnybrook School. Every day, even on Sunday, I'd be up at 5 a.m. doing my regular chores: milking the cows and chopping wood was a start. I'd take time to brush the horses until I heard mom call me in for breakfast. By 7 a.m., Dad would be out clearing brush on our property, and my sisters and I would be off on our hike to school. We really did walk a mile, which seemed pretty short then.

School was a fun place to be when I was five and with Anne and Mary in the same class I was under good behavior for the most part. But I can remember embarrassing the hell out of them the day when our teacher, Mr. Gary, asked us each to share with the class the place where we were born. Most kids may have mentioned hospitals — but I said, "My mother had me in the corner of the chicken shack". Mary and Anne went red in the face, but what I said was true, sort of. What became the chicken house was my parent's first home. Eventually, Dad built us a bigger house, and we left the chickens behind.

Mr. Gary made quite an impression on me. There were forty-five kids in the classroom, grades one through twelve, and he had to teach them all. It's understandable that he had to be heavy with the discipline. He ruled that class with an iron fist. A poke to the back of the head with his chalk-board stick was Mr. Gary's remedy for a loose mouth. And you never wanted to get caught talking or acting up when his back was turned. His solution was a quick chalk brush whipped at your head. I learned to duck fast and it would get the kid behind me. I was on good behavior with Mr. Gary for the most part but one day, during recess, the boys dared me to push this girl, Elizabeth, into a nearby mud puddle. Of course I did. Mr. Gary was going to give me the strap but three or four of the bigger students, who dared me to do it in the first place, stood up for me. I never did get the strap.

During the summer months we'd go to church. I'd usually be up early on Sunday anyway, chopping wood so that we'd always be stocked up a year ahead.

By the time I was eight I had a few good friends. Alan Abar was a good looking kid, with dark eyes and dark brown hair, and George Paisley was big in the shoulders, like a buffalo. The three of us were left handed, and Mr. Gary did his best to train us to write with our right. One of our favorite activities was catching gophers. We'd flood their little holes with a bucket of water and as soon as their heads appeared, we'd conk 'em with a stick. Their tails were worth a cent a piece in town.

Another regular activity was shooting squirrels. A box of fifty .22 shells cost 15 cents, and I'd usually pop off about forty-eight or forty-nine critters per box. I'd sell the squirrel's hides in town for 6 cents a piece.

Lloyd Chiddick was another friend of mine, and he was a tough kid. Like his father, he was a scrapper. One time Lloyd went moose hunting with my dad and his. I never knew my dad to buy more than two shells at a time and that was all he'd ever take out with him. He probably couldn't afford a whole box of them.

The three of them were pretty far out in the bush when Dad came across a moose head, a discard another hunter had left behind. He knew Lloyd was coming through the bush so Dad kind of jammed the moose head between a couple of trees, horns and all. Poor Lloyd. I know he had quite a few shells and, well, he emptied his gun on that thing. We teased him about that one for quite a while.

There was a friendly trapper named Lyle Cleveland who lived on a farm about a mile away from ours and he was quite a character. He must have been in his late forties then. I'd hurry to finish my daily chores at home then race over to Lyle's to watch him skin animals. He steered his sleigh with a dog team of eight and would go out maybe fifty or sixty miles trapping along the Saskatchewan River. Mostly busy in the winter months, Lyle skinned coyotes, foxes, and the odd mink. Man, he was fast, and it was the first time I had seen anything like that. I never did skin with him, I just stood quietly and watched. I got Dad to teach me to strip tails real quick. We'd take them to town and the municipality paid us a penny a piece for them. There were two different kinds of gophers — the grey gopher, in the south part of the province, and in the north there was the striped gopher. It looked a bit like a chipmunk only bigger with a bushy tail. The animal tails were of no use to the municipality. They just threw them away and burned them. It was merely a method used to keep the little rodents from eating valuable grain. I invented a trap out of a tin can which made it easy for me to catch them. I cut a little star-like hole in the side of the can and baited it with a little bit of meat. Squirrels were easiest to catch; weasels were a little tougher on account of their necks being almost bigger than their heads.

My opening had to be just the right size to hold them in there. Eventually with trapping money I was able to buy my own school clothes and books. And with a bit more effort, I made money from picking seneca root, which the Indians used to make laxatives with. I could get ten cents a pound, which was a good amount of money. It was rare to find a really good patch of seneca, but if you did, you could get two or three pounds out of it.

With my earnings I bought practical clothes; dark britches and putties — a type of legging I'd wrap around my calves to protect me from the snow. My blond curly hair would get unruly and I found a bit of axle grease worked well to control it. When that didn't work anymore, Dad would just take the shears to it. Anybody who needed a haircut went to see him.

Sometime around the end of June, Dad headed out on what turned out to be a very short hunting expedition. He returned just days later in what was left of his underwear, and the remnants of his belt gently looped around the neck of a small bear cub. He'd shot the old sow bear and the cub ran up a tree. Dad knew he could get $10 for a cub at the museum in Regina, so he did his best to drag it home. He tied it up in the back of his shop and within a day it had buried everything it could find. We kept the cub around for six or eight weeks before we finally sold it and it was my job to feed it. I liked having the little guy around, but I knew better than to get attached to animals.

Even though we didn't have much money, mother always made sure that we ate well. After our first season in Sunnybrook her garden was going strong. Mother grew everything, but mostly carrots, turnips, parsnips, and potatoes. The root cellar would store our crops until spring. Nothing froze down there and with winters that sank below fifty degrees, food storage was usually the biggest problem. Frost would usually spread four feet or so into the ground, so the cellar had to be deep. Dad organized it with little wooden bins, and he'd use poles to separate the vegetables from each other. With the flat prairie land laying open to the exposure of drifting snow, winters were pretty harsh. But we were prepared; we had to be. We always had three or four hundred pounds of flour stored in the attic and lots of water on hand, because it wasn't uncommon to be snowed in for weeks.

In 1934, we said "goodbye" to Sunnybrook. I was nine years old. Dad felt there was too much debt against the quarter section piece of land we lived on and that we'd have better opportunities elsewhere. We remained in Nipawin and moved into a house which belonged to an Indian named Lucier. It was a small house located up the bank on the Saskatchewan River. He allowed us to live there until Dad finished building our next home.

That summer at Lucier's was a good one. My mother and sisters spent afternoons swimming in the Saskatchewan River, I chose to stand by and watch. I never did like water much; it scared me then, and it still does. But not Mother. She wasn't afraid of much, and she was an excellent swimmer. She'd have no problem with swimming to the other side of the river, which was about a half a mile across. Us kids knew to stay between the bank and the sandbars, where it was shallow and calm. That river had quite a current and the sandbars shifted quickly. We knew to be careful.

Our family friend, George Nixon, lived across the river and once in a while he'd get his horse into the water, grab its tail and get towed across. It was a fun time. We'd also go down to that river and catch fish, which Mother would smoke. She canned as well, with some old jars she had collected over the years. They were happy days.

3

Forest Trail

For the next few winters, we remained in Lucier's house, and Anne and I attended Forest Trail School. It was about three miles from the house on the river. Mary had finished school by then. It was just Anne and I, driving the dogs and sled to school, and all the way Anne would be shouting "Mush! Mush!"

A lot of the things Dad told me as a boy have stuck with me. But one of his most influential statements was, "Bud, if you haven't got respect for women and age, you haven't got a hell of a lot of place in this world."

Good advice, at a young age. My teacher, Miss Black, was an older lady and she had a hell of a time disciplining the kids. The ages varied so much and because there were so many Polish immigrants, a lot of fourteen and fifteen year olds couldn't speak English. I imagine it was a real tough job teaching then and teacher's wages were only about thirty dollars a month. There was the odd kid who would take a kick at Miss Black when she was giving them the strap, but not very often. I never did get the strap but that's not to say that I was always good. I'd be caught out of line once in a while for talking, or pulling a girl's hair, or maybe locking someone in the outhouse. She had different methods of handling different kids. For example, she'd make me stand at the front of the school with a couple of books in each hand and I'd have to stand there for a long time with my arms straight out. Or else she'd draw a circle on the black board and make me stand there with my nose in it. I tried to not let that happen too often.

Anne and I picked up an after school job doing janitor work around the school. For $3 a month I'd get the fires going and pack water in. Anne would sweep and clean the black boards, which only took about an hour. By the time spring came, the outhouses would be getting pretty ripe. The trustees would decide they'd have to be cleaned. In the last month of cold weather someone would be hired, for a couple of dollars, to beat the growing mound down, then jump down the hole with a bucket and shovel it all out. No volunteers for that job.

Toilet paper was a rarity then and my family was no more privileged than the next. It wasn't until a couple of years later, when we had a teacher boarding with us, that mom actually bought some. Before that we used the Winnipeg Free Press, or if we were lucky, the Eaton's catalogue. They always had great hardware and harness sections to browse through, and those pages were smooth. We'd be well into the harness section come spring.

By the time I was ten daily chores were more than just milking cows and brushing horses. I was pretty comfortable with castrating calves. I had caught on to Dad's techniques. A good sharp knife and a quick cut to make two little slits in each bag; a couple of stones in your hand, you knew you had a steer.

There were plenty of times in the winter when Anne and I would come home from school and have to melt snow for the animals because the well was frozen. I'd build a fire under a barrel, throw a bunch of ice in it, and we'd have water in no time. Mind you, when the cattle got warm water, they drank a hell of a lot more, too.

Clearing brush and burning it all day was another responsibility, and I'd be blacker than the inside of a cow by the end of the day. We worked well together, Dad and I. I was smoking the odd cigarette with him then. He would roll me one at the table but he made it clear if he ever caught me taking any tobacco without his permission he'd tan my ass.

An exciting event happened once every couple of months when we'd all jump into the wagon and head to town. We looked forward to it so much that my sisters and I prepared for the event two or three days in advance. I'd wrap my squirrel skins up to sell. We didn't have many clothes to choose from but what town clothes we had were laid out. The team and sleigh we'd have prepared the night before.

Nipawin was small. It had three grocery stores and a couple of butcher shops, and I'd use the money from sales on my animal tails to treat myself to a new box of shells which I used sparingly. Usually, we'd bring one of our two hundred pound butcher pigs along and trade it for staple items: sugar, salt, pepper and coffee. We never bought soap or toothpaste. Mother made those from scratch with lard and soda.

In town, the grocers would always give us kids a big paper bag of candy; a mixture of gum and jelly beans. Mary and Anne and I always looked forward to digging through the groceries, and there wasn't a helluva lot to dig through. But we knew we'd find that bag of candy, and it would hold us off, wired, for another month or so.

Mother sure loved her pigs. Gestation period for pigs is four months and sometimes there would be up to three litters in a year. There were always about forty or fifty of the big beasts wandering around. They were happy to roam in their pasture and the more they rooted out there the less feed they needed. It was Anne's and my responsibility to keep the garden weed-free. When we'd completed our assigned row, the lot would be taken to the pigs and they'd gulp it all down. For some reason, I never liked pigs after that.

4

Home Sweet Homestead

It was called a homestead because you had to pay $10 to the government to lay claim on one hundred-sixty acres of land. We looked at it in terms of the government betting $10 on 160 acres of land, that if you could survive on it for five years you'd get the title, providing a certain number of improvements were completed. The land officer would pop out for surprise inspections to ensure we were keeping up our end of the deal. By the end of the third year, we were expected to have a house and barn built, as well as three or four acres of land cleared. And by the time the fifth year rolled around, they expected to see more land cleared, with livestock. If all of these improvements were completed, the title was yours.

We moved into our homestead in 1936, when I was eleven. The two story house seemed huge to us at the time, although it only measured 24' by 28'. Dad had to haul the logs quite a distance (twelve or fifteen miles), because there weren't any decent building logs nearby. It took about sixty logs or so to build that place and with the help of a few neighbours it was finished within a year, dove-tailed corners and all. I was pretty proud that I'd helped shingle it.

The homestead didn't have any insulation and there wasn't

this tongue and groove stuff for flooring — just plain boards. And you hoped to hell that when the boards dried they wouldn't crack. For the first few years, the logs on the outside were mudded until Dad stripped them to put another layer on. It was a family event. We'd jump into that tub of mud with our bare feet and mix the chaff into a finer paste. With a thick layer of mud on the outside walls the homestead stayed really warm. Inside it was a beautiful house, with a kitchen, dining room, and sitting room, and mother and dad's room on the main floor. Upstairs was a separate bedroom for each of us, as well as an extra room for storage. Our furniture consisted of a dining room suite that mom had brought from Brooking and a few chairs. Our mattresses we made ourselves out of straw.

It was around that time that I fell my first tree. I was so proud. It measured a foot across and sixty or seventy feet tall, and it took all day to chop down. I stood there and watched it fall with no clue as to where it would land. That was my version of playing.

Back then it was rare to hear the sound of an airplane let alone see one. If we heard the engine's soft hum in the sky everyone would run out to try and get a look.

When I was twelve, I decided to quit school. Not necessarily because I didn't enjoy it; I liked the basic subjects (nature and science, health, math, and spelling). My favorite subjects were history and geography which Miss Black taught well. But it seemed more important to me to go out and start earning a living. I have always told people that I finished half of my twelve, which is true, sort of. I finished up to grade six. I guess I figured I'd learned enough in the classroom and it was time to go make some money. I enjoyed working around our homestead and Dad had always said, "If you expect to get a man's pay, you've got to do a man's work." I'd be damned if I was going to work for half wages just because I was a kid.

For the remainder of that summer I kept myself pretty busy around the homestead. I thrashed grain for a couple of months in the fall then worked haying and making railroad ties when thrashing season died down. Haying took place about four or five miles from home at the swamp meadows. Sometimes we'd be out there for days, building up loads for the cattle to haul home. If we got four or five days of dry weather we had to take advantage of it. We'd sleep under the trees at night and wouldn't go home until we had enough to fill a rack.

The year after, I got a job haying for an Irish man named Al Morrisson. He was fifty-one years old, chewed tobacco, and loved my dad's green, home-made beer. Al and I would sit down together and drink a bottle or two of the stuff before he'd help me load up my hay. I'd have a helluva time getting my load on because it had to be thrown up so high. He was a funny character, directing his big old hereford bull and saddle horse around the field. Christ, it looked funny.

That year I gained experience running alfalfa and grain combines. When the ground froze, the work was especially tough. Tractors were built on steel wheels and the motion over the frozen ground was pretty bad. Every time one of those lugs rolled around I felt like my teeth were gonna fall out. I'd wrap myself up in old potato sacks... anything to keep me warm. And all day long I'd sip buttermilk, until it turned into butter!

We didn't have any neighbours then (most houses were at least a couple of miles apart), but I befriended a giant of a man named Weldon Cockrel. He was 7' 2", and lived alone on his fox farm. Weldon hated women. If anything went wrong, if the weather was bad, it was always a woman's fault. He didn't have many friends (most people were married), and it took a lot to get a smile on his face. But he didn't mind me. I knew he wasn't all bad. And he was a good housekeeper, for a bachelor.

One day I headed over to his house with an old 41 Swiss gun I owned. I wanted to get rid of it and Weldon used to shoot horses to feed his foxes and mink, so I knew he could use it. I traded him the gun for a windup gramophone. I'd wind it up and listen to my favorites: Wilf Carter and Hank Williams. Weldon had a violin and he could play that thing like a house on fire. But eventually, he made a career change. He got rid of his fox farm and became a grain farmer. I still have that old gramophone.

Not long after my oldest sister Mary turned twenty-two, she married. And about a year later, Mother delivered Mary's first baby. I was riding my horse around outside and I could hear Mary screaming away inside the house. Mother delivered a healthy baby girl which they named Nina. I took her for her first horse ride when she was eight days old.

As with most things, I'd learned a lot of tricks from watching my dad. Shoeing horses was one of them. "Tie 'em up and let 'em fight the rope instead of you," he'd say. By the time I was thirteen, I was confident enough to do my own. Unfortunately, shoeing also led to my first experience in learning that you can't trust everyone. I'd made a deal to shoe horses for a local logging camp. At night, I worked by the light of a lantern for $2 a horse. The owner didn't pay me a damn cent. From that day on I was turned off of working for anyone else.

Life could be hard then but we didn't dwell on it. We just dealt with it. It was all about survival. And memories of dancing to Dad's melodious mouth music still linger with me. A carefree evening of dancing to his mouth harp took place in our family room at least once a week. I'd dance around with

Mother and my sisters, until they got mad at me for trampling all over their feet. I learned some good moves, but Anne's were always the best. She was a natural.

Aside from playing a mean mouth organ, Dad was an excellent artist. He was especially good at drawing pictures. He'd draw pheasants and rose bushes on old grain bags and with a hook and some old rags mother would make them into hook rugs. She'd sell the odd one for a dollar as a wall hanging or for draping over tables, and I would lay them on the floor around my bed to keep my feet warm on those cold Nipawin mornings.

Considering the homestead wasn't insulated, it wasn't all that cold during the winters. Sometimes, by morning, ice would have formed on the washbasin, but that was rare. Dad built a stove out of a gas drum which he'd fill up at night and it would last through until morning.

In the summer time we played horseshoes and baseball. Using an old winter mitt as a glove, or using just our hands, we found fun for free. Come winter, the game was hockey. The goalie would throw on a big sweat pad for protection, and with a crooked willow stick and a frozen horse turd as a puck, we had our own NHL.

When the games ran out there was plenty to keep us busy on the farm. Like the chicken house Anne and I built. It measured 12' x 14', and it took us at least a couple of months to complete. Dad helped me chop down trees for the wood and showed me how to use a string to make sure everything was square. Other than that, Anne and I did it all on our own. Anne was quite concerned about how much space I'd left in the front of the house for a window. Not that we had any glass to put in there, just plastic. We nailed up whatever we could find for the rest of it. It became home to about three dozen chickens and sheltered them through freezing winters. When it got really bad, we'd light a little wood stove to keep them warm because it wasn't uncommon to find the damn things with their crowns or toes frozen right off. Chicken frostbite. The survivors would start laying in April and in the fall when they started to molt we'd collect the eggs for Mother to store. By the time the chickens had finished laying for good we'd have twenty or thirty dozen eggs stored in big, glass, water units down in the cellar. There they would keep all winter and would mainly be for baking. The cellar wasn't our only method of refrigeration. By hauling chunks of ice from the river, which we'd cut up using a handsaw, and burying them in sawdust in the ground, we'd have ourselves a functioning freezer throughout the summer. That method took about a week to prepare and if we had the time, we'd build a storage shed over top of it; a veritable walk-in.

Looking back, my sisters and I always got along well. The whole family did. We worked together as a team to get things done, to improve the homestead, to improve life. It was a pride thing. I think it's a shame that through the years, as a society we've lost a lot of that. That's progress, I guess.

5

Making Moonshine

The recipe was a family secret, and we had it down. We'd throw some wheat in a forty-five gallon barrel, add one hundred pounds of sugar, a box of square yeast cakes, and ten pounds of grapes or raisins, if we had them. We'd leave it sitting at room temperature, preferably in a place not visible to anyone, and store it upstairs until it was ready. Some people would bury barrels of the stuff in a manure pile — the heat would help in the process. Within months, we'd strain off the sediment and use a coiled copper tube to pass the moonshine down through charcoal, to filter out the fossil oil. Dad would usually make a couple of gallons at a time.

There were pretty serious fines for making moonshine, like three to six months in jail, unless you could pay the bail. Which, at a couple hundred dollars, wasn't too likely for the average working man. And at every schoolhouse dance there'd be a bootlegger trying to sell the stuff. It was bad news to get caught with anything having to do with moonshine. Dad would put a cork in each end of his well used copper worm, and bury it in a manure pile. He never did get caught by the authorities but there was a close call once when they came out with their four foot rods, poking them in the grain bins, looking for evidence. Dad was nervous but he'd hidden his evidence well.

I never needed to sneak moonshine. From the time I was eight or nine years old, Dad and I would have a little shot of the stuff before I went out to do my chores. It felt like fire going down my throat and I liked it. If it was double distilled, it was even stronger. Testing a batch of moonshine was simple; pour a bit on a spoon and light it. When you tipped it, if it burned a hole through the floor and there are only a few drops left on the spoon, you knew it was a good batch. And once mastered, moonshine never gives a hangover.

Of course, not everyone was honest in their making. Some would add Gillis lye as filler. Lye is the solution you get from leaching wood ashes, consisting mostly of potassium carbonate, and meant to be used in the making of soap and glass. Terrible stuff. Bootleggers would sell a beer bottle full of lye-tainted moonshine for 50 cents. One night five of us, including Sam Delahay and an older fellow named Charlie Bleich, all chipped in 10 cents for a bottle of the stuff. That familiar fire feeling in my throat was there but something wasn't right about the flavour of it. Thanks to Dad, I knew what a good batch should taste like. He'd warned me to be leery of a bad mix.

The bottle went around the circle once but when it got to me I spit it out and said, "It's no good boys. Don't drink it."

The others had only had a bit, except for Charlie. He'd taken two or three good shots and it blinded him for about two or three months, until it finally killed him.

Dad would give me a little vanilla bottle of moonshine in the morning to take to work with me when the days were cold. Moonshine was especially mandatory when I went to the dentist. Homemade pain killer. I had a helluva toothache one time so I borrowed a young preacher's bike and rode it into town. I slammed that little bottle of moonshine before I went into the dentist's office because there was no such thing as freezing then. The waiting room was never full. You were in and out in no time. With no anesthetic, it never took too long for the dentist to get the job done.

We avoided going to the dentist by making our own homemade fillings. I'd burn magazine paper in a can on the stove and it would leave behind a waxy residue. As it cooled it would become real sticky and I could apply it with a little stick or something. That would sure stop the cavity for a while, at least until it wore out. I can't remember when I last sat in a dentist's chair. A good pair of pliers still gets the job done.

6

Fire Fighting

In 1937, I was offered a job fire fighting in the Torch River area. I'd never done it before and I was pretty excited at the prospect of making ten cents an hour. It was a warm night in April when I got into town to meet up with the other fifty guys who were going. I bought a pair of work boots and borrowed two old horse blankets from the local livery barn. With boots and bedroll, I was set.

The journey went on until after midnight, with the group of us taking turns riding in the back of a truck. We finished the trek with a team and wagon. It was a bumpy ride until about 4 a.m. when we finally reached the site of the fire.

We were taught the routine technique in fire control. With our shovels we dug ditches and trenches, then lit a backfire and let the main fire meet that fire before we were able to put it out. When the two fires met we were careful to catch jumping sparks before they turned into spot fires. With a backpack, a hose, and lots of manpower, things never got too out of hand.

Days were eighteen hours long and the first fire took us about six weeks to put out. My sister Anne had married the man who was our foreman, Joe Peterson. Standing 6' 1", he was a big guy with a heart of gold and he looked out for me.

For the most part, Joe was an even-tempered man but he could be mean. And when he got angry just his voice was enough to scare anyone. Joe broke me in as a power faller and I looked up to him. I especially admired his mustache. It wasn't much more than a thin line of hair lining his upper lip, but I wanted one. I didn't quite have the ability to grow my own then so I'd draw one on with a piece of charcoal. I still have a little 'stache like his to this day.

On my second fire fighting excursion we camped along the Saskatchewan River. There were one hundred and fifty guys on that job and, at the mere age of fourteen, I was just a kid in the crowd. We slept under the stars, or whatever tree was still standing.

Fires had chased most of the game out of the area but one night as I lay sleeping I awoke suddenly to the feeling of something soft and warm sniffing my nose. As I laid there in the pitch black night, too scared to move or say anything, I knew it was a bear. I kept my eyes shut, focused on my slow breathing, and then it was gone. The next morning I told the guys about it and they told me I was crazy. But low and behold, we found tracks left by a mama bear and her cubs all around my blanket. Nobody bugged me much after that.

On that same job, I befriended a man named Clarence Lindman. He was a year older than me and stood about 6' 4". Everyday, the group of us would pile into about six or seven row boats and paddle across the Saskatchewan River to our destination. One kid was just scared shitless of the water, some sort of phobia, and he insisted he wouldn't cross that river unless he was unconscious. So, twice a day Clarence would knock that kid out just to get him across the river. One day Clarence stood over the kid's body and said, "Crazy what we'll do for 10 cents an hour!" I had to agree.

I met up with Clarence years later, after he'd returned from the army and earned a police badge. That same night the Nipawin Bank got robbed. I guess he should have been available, but we were off pit lamping (night hunting) which wasn't exactly legal. When he found out about the crime he fled from my house in such a flurry, he left his police-hat behind. It was just as well. There was so much blood on it from hunting, he would have been busted for sure. For a while I wore it around town for laughs. He never did ask for it back.

The fine for hunting out of season was pretty steep but people risked it anyway. The game warden was a friend of my family and used to stop at our place before he'd head out to Raven District where he lived. One night on his way out there he met up with our neighbour, Mr. Salsbury. Noticing the large load in the back of his truck, the game warden said jokingly, "By God Salsbury, you'd better get that load of wood home before it bleeds to death."

The warden was usually pretty lenient when it came to giving out fines, but I remember he did take Dad's gun away from him once for a year, for shooting moose out of season.

7

The Sawmill Adventures

By the time I had turned seventeen, I contracted myself out to a local sawmill skidding logs. Working four horses for sometimes twelve hours a day, I was averaging one hundred logs a day. At 7 cents a log, I wasn't doing too badly. But I guess they figured that was too much money for a kid to be making. I took my last check at the end of the second month, and went hewing ties for the rest of that winter.

But once again my ability to work harder than the others caused conflict. I was faster than the older, more experienced guys, and I was earning a lot more than them, too. We got our paychecks a couple of days before Christmas and mine just wasn't right. I'd earned less than half of what I'd made before. I knew I was a hard worker and I didn't like being cheated, but I also decided that arguing over it wasn't worth my trouble. That night I loaded up my horse and gear and headed home. The weather was fifty below and Nipawin was fifty-six miles away. I rode through the night until I came across an old barn which I stayed in until morning. I was so discouraged and angry from being cheated on my pay check that the next day, I vowed to never work for anyone else again. I bought myself a sawmill and went into business. It was a bit of a struggle getting that little sawmill home. The easiest route was down the frozen river and it took Dad and I, with two teams of horses, all day to drag it home. For fifteen or twenty miles down the icy river we struggled, slip-sliding with the horses, trying to make it home before dark. We had a close call when Dad's team stopped short because of a huge hole in the ice. With mill intact, we coaxed the horses around the frozen crater and led them home.

Once we had the mill set up business started pretty quick. Word traveled fast and soon I was busier than a cat covering shit on a tin roof. Dad and I worked like mad. With an axe and swede saw we'd fall the tree, then hew it before cutting the top off at eight feet to haul it home. It wasn't too long before we had a steady stream of clients and by spring I was able to upgrade to a larger mill. I bought it for $200, along with a 1530 McCormick Deering tractor which I used as a generator.

I was averaging a tie a minute, making about six hundred ties a day. And at 10 cents a tie, I was finally making some money. It felt good and I know Dad was proud of me.

For the most part he and I got along well. I can really only remember one altercation between us which happened when we were sawing ties. I must have already been a workaholic by then, because I had worked steady with the sawmill for three or four months. Dad was stacking the ties behind me, which was the heaviest part of the work, and we were going great guns. I know I had quite a temper and I'd fly off the handle if something went wrong. This one day I lost my temper. I had thrown my hat on the ground and was jumping on it. Dad had had enough. He grabbed me by the throat, brought me eyeball to eyeball with him and said, "Goddamnit! If you're gonna work like a man, be one!"

I sat down pretty quick. I never even answered him back; I

felt so damned embarrassed! It straightened me right up. He had a tendency to do that. That was the closest thing to an argument we ever had and it did me a world of good. I didn't fly off the handle quite as often after that, especially if Dad was around.

In 1939, I was thrashing with an old fellow named George Whittaker, a World War I veteran. He and I were sleeping up in the loft when we heard that the war had broken out. Warsaw and Poland had been struck, and I can remember the local immigrants were pretty sad. But old George wasn't affected by any of it. And I'd come to realize that he wasn't exactly sane. One night he pulled a knife on me. "By God, Buddy," he said, in his raspy old voice, "this is what I use." He pulled out a big hunting knife and I remember a crazy gleam in his eye. "This is what I use as an equalizer if nobody listens to me." He scared the hell out of me.

But then he looked out for me too. Early mornings when it was time for me to get up and tend to the stock, he'd let me sleep in. "You sleep and I'll feed your horses for you," George would say. I guess it said a lot about him, considering they had to be fed by about 4 a.m. I learned at a very young age, feed your animals, then feed yourself.

It was a regular occurrence to witness one of George's crazy spells. With his gun and an attitude, he'd walk forty or fifty miles to the next town, looking for the Germans. At one point he even dug a trench around his shack and patrolled it all night long, in case the Germans came. The last thing I heard about old George, his stud horse pissed him off so he shot it and buried it right in front of his door. "I'll watch you now, you bugger," he said.

There were a few crazy characters in my family's circle of friends. Sam Delahay was another one. I can remember one summer, during milking time, he had corralled the cows and was making a smudge — a fire made of wet grass or manure, or whatever you could find to make smoke to keep the flies away. Sam wasn't just crazy. He was a mean bugger, too.

He'd be milking away and this little kitten would come around looking for warm milk. Sam liked to grab the cat by the back of the neck and shove its head in a fresh cow pie. He had a demented sense of humor. One day, it was raining so hard that we couldn't do any thrashing. Out of boredom, Sam decided to have some fun with calves. He took a rope and rubbed it under that calf's tail until it was raw, then took a bottle of turpentine and poured it in there. Well that poor calf went crazy. The old guy we were working for came out and said, "Now that's the way I like to see them calves. They're healthy and happy when they're bucking around like that." He would have killed us if he'd known what we'd been up to.

I recall the damnedest thing happening that same year, when old Mr. Stacey went to the outhouse and didn't return. Eventually his son Norm went out to see what was up. He found the poor old guy deader than a door nail sitting on the 'john'. Mr. Stacey's other son, who was in his early forties and extremely overweight, came up for the funeral. The casket was being lowered into the ground when suddenly the son had a heart attack. He fell right down the hole on-top the casket. Died right there. It was a really sad time and I guess because of stress, Mr. Stacey's wife died soon after.

I became good friends with the youngest son, Norman Stacey, but he didn't have the best of luck either. One day when he was working the pea combine, he cut his index finger right off. It came out clean and landed up on top of the grain. His wife, who was a trained nurse, grabbed the finger and her husband and got them to the hospital. She just couldn't understand why they couldn't just glue that finger back on.

In the fall of 1940 I made money 'stooking': preparing grain bundles for thrashing time. I was only earning $3 a day, but because I worked so hard the old fellow I was working for paid me enough for three days' wages in one day. I hauled whatever I could, grain in the fall and logging on the side. I was always doing something. I had learned early in life that I was usually better off working for myself.

8

Sawing Ties

In 1942 I signed up for the army, but they turned me down because of a flutter in my heart. I continued to find work on my own and worked on and off with Sam Delahay. Around the same time other people were signing up to join the army, I hired Sam to drive a team of horses about eighty miles to get another horse from his brother-in-law. A couple of weeks passed and there was still no sign of Sam. Well I didn't know what to think. Anything could have happened to the guy. It was about three weeks later, on a trip into town that I found out Sam had left that horse and the team in the livery barn and joined the army. It didn't surprise me when I heard he was in detention most of the time.

At seventeen I went cutting ties which was really how I got my start in life. I lived in a fire ranger's cabin and I'd stay there for a week or ten days, until I ran out of feed for myself and for the horses. When that time came I'd usually return to the homestead. I remember one moonlit night, I'd been walking about five or six miles when the timber wolves started following me. I couldn't see them, but I knew they were there. I could feel their eyes watching me. I can't remember ever being afraid of them, as timber wolves seldom ever attack. And, of course, I had my Dad's voice in my head, "Don't be scared kid, just keep walking."

I never got lonely up in the cabin, I liked being on my own. And there was always an animal about. In the spring I caught a little groundhog, which I'd feed pancakes to, and a little blue-jay was my friend, too. I never worried about my safety, being alone in the middle of the woods, but one day when I returned to the cabin, I realized somebody had been in it. The cupboards had been rummaged through and there were pots and things scattered around. It happened a few times before I realized it was draft-dodgers looking for food. I shot the odd squirrel, but if I really needed meat, I'd meet my basic needs by simply going out and shooting a deer.

Generally, I spent the daylight hours making ties. I was averaging about six hundred ties a day. Once I'd peeled all the bark off, I'd load the ties up to the team of horses to haul to the railroad for pickup. They were then graded by size and class. 7' x 9' were 79 cents a tie, 6' x 8', 59 cents, and 5' x 7' were 45 cents each. Dad used to make big ugly draw knives for peeling the bark off and because they were so sharp they were especially handy for chopping off the big knots. You had to be careful you didn't cut your belly right open with one of his knives. I'd strap a chunk of cowhide around my waist to protect myself. If you did cut yourself, there was no rushing to the hospital or anything. You had to deal with it on your own. And for small cuts, we used a natural disinfectant — pee. Dad taught me that when I was probably seven or eight when I cut my hand up with some rusty old fence wire. "Don't pick it up with your hands," he'd said, "use a stick or something."

But of course I used my hands and before I knew it, I was bleeding. Dad took a look at the wound and said, "Pee on it."

I didn't question him. I just did it.

Through those days and into the nights, Dad and I sawed

ties. Then loaded them into a truck we'd hired for transport to the Nipawin Railroad track. The tie inspector would do his job before the three or four hired hands would load the ties up the skinny planks we used for a ramp. It was heavy work and we barely stopped for lunch, but I didn't need much during the day to keep me going. I'd be happy without a lunch break as long as I'd had a good breakfast and supper. You could wring the water out of my pant cuffs I sweated so much. My main fuel was buttermilk. I sipped on that stuff throughout our usual twelve hour day. By two in the afternoon, even the good guys were played out. They would have put in a helluva days' work, but that buttermilk kept me going strong.

In 1944 the NDP were elected in, and everything changed. To me, NDP always stood for "Naturally Dumb People." They changed the rules of the whole natural resources industry for the worse. Fishing and mining all went downhill. The locals used to go up and haul fish out of the northern lakes then peddle them door to door. After the CCF got in (Canada's Crazy Fools), the fish were sold only to them. The fishermen got three cents a pound and if they wanted to take the fish home they had to pay the government nine cents a pound!

Similar restrictions began happening in the forestry industry as well. In the first year, we were given the same profit-cut as the year before. The second year, we got half. The following year, the government took it all. Eventually, they had shut it all down. Out of the sixty-three mills that had existed locally, mine included, there was one remaining and it belonged to the CCF. Where we once horse logged and took care in not damaging the younger trees, the CCF took their cats in and wrecked it all. Basically, anybody with any ambition had to leave, including me. I walked away from my sawmill and along with most others, eventually made a move to find other work.

9

A Bit of Social Life

The year was 1943 and I was sixteen. Dad had taught me to shave using no more than a straight razor and a bit of his soap-in-a-mug, around that same time I bought my first truck. When I wasn't shoeing horses I'd drive that green 1923 Model T around. I purchased her for $50. With no gas pedal and only a hand shifter (like on a tractor), I cruised around, barely able to see above the dash. Sometimes, for kicks, I'd steer from the passenger's side and people would think there wasn't anyone driving. The surprised look on their faces as they ran like hell to get out of the way was pretty amusing. I didn't worry too much about getting caught with such a prank; there was only one mounted police in the area and he was never around.

I drove that truck for a year or so before I wrecked it, but it wasn't long before I traded my horse for a Model A Roadster. I cut the back seat out and created a sort of box to haul small livestock around. It was cheap to run, especially on purple fuel. It didn't need much of that 'marked gas'. And considering a forty-five gallon barrel of the stuff cost $7, it was damn cheap to run. After about two or three years, I traded the roadster in for a little stud horse.

My buddies and I used to have a lot of fun taking the town

---

girls to the dances. We'd bring along a little bottle of whiskey and if things went our way, we'd get to put a hand on her knee.

I thought I was a veteran at dating by the time I was seventeen, and there were some girls who stood out more than others. Jean Wolfe who worked in the local linen store was sweet, but I was madly in love with Mary Thompson. She lived quite a ways across the Torch River. One time I took her to the movies, which in those days only cost a dollar. We rode in with my team of horses and a load of wood that I was bringing to town. After the show I drove her home and we smooched for a while at her doorstep. When 3 a.m. rolled around and I still wasn't home, Mother and Dad were worried, so Dad saddled up his horse and came looking for me. He said that when he found me, I wasn't too far from home. I was sound asleep behind the reins — the horses were going home on their own.

Probably one of my most embarrassing dating moments was with Mary. I jumped in my 1928 Chev one Sunday afternoon and headed to her house. Cruising down the gravel road I slowed down to take a corner and the damn thing stalled and just wouldn't start up again. I was starting to panic a bit because it wouldn't look good being late for our date. I got out of the car and as I peered under the hood I realized the problem; the battery was gone. It had fallen out of its place about a half a mile back down the road. I raced back, grabbed it, and hurried back to the truck to set it under the hood. As I drove away, I didn't think much of the small patch of battery acid which had leaked onto my pants.

When I finally got to Mary's house she and her mother were headed to church. I decided to join them.

It was nice sitting on the bench beside Mary but I was a bit nervous. At least we didn't have to talk. It wasn't until I noticed that my pants were starting to fall apart from the battery acid, and I saw my boot laces laying in little bits on the floor that I felt like an ass. I guess Mary had a pretty good sense of humor because she laughed hard at that. We got the heck out of there and I headed home to change. I was happy that she still wanted to see me after that.

But of all the girls, there was one who was on my mind from the moment I laid eyes on her. Her name was Phyllis Dube. She and her family had moved from Poland in 1929, and she was eight years old the first time I met her at school. In fact, I was thrashing at her house at around the same time the war broke out in Warsaw, in 1939. Phyllis was a tall, slim girl, with shiny brown hair and she had the nicest legs. I remember our first date clearly. It was February 18, 1945, and I took her to the community hall dance. We danced all night until I had to take her home.

In 1947, my folks were contemplating a move out to B.C. My sister Anne and her husband Joe had moved there six years earlier, and Mother and Dad decided to do the same. They made arrangements to auction off the homestead's contents. I had purchased one hundred and sixty acres of land for $800, and had kept myself busy logging the timber on it. But work was scarce again and I started to think that a move to the coast was a good plan. I traded in the land and my sawmill for a second truck.

My folks' auction took place on a frosty November morning and there were close to three hundred people attending. The cows and horses were brought inside the barn so that people could get a good view of them. Outside, there were tables loaded up with everything you can imagine: from jars of canned pickles and picture frames, to axes and adzes. The hay wagons were loaded with stuff, too. I knew that I felt relieved to be leaving Saskatchewan, but I was sad to see all of our belongings go. I was especially sad when my horses sold. My buddies and I sipped moonshine and listened to the auctioneer. His voice carried on into the late, crisp evening air. His technique must have been effective because we sold close to everything. The only things I can't recall being accounted for are dad's blacksmithing tools. It's possible that he sold them privately. Then again, he may have just given the items away. In the end, the auction brought in about $2,800.

I hadn't seen Phyllis for a while but she was still on my mind and that night I bet this old Polish shoemaker a case of whiskey that he couldn't convince her parents to let her marry me. She was only sixteen and I didn't think I had a hope in hell. As it turned out, I lost the bet but won the girl.

Whenever there was a wedding planned the first priority was usually to hurry home and make enough moonshine for the festivities. We knew it would take at least ten days before it would be ready. Phyllis's family was Catholic and there wasn't a Catholic church in Forest Trail so I made a deal with the priest to marry us in his own house. He did, but I think that in his mind we were never truly married.

Phyllis and I both passed the mandatory matrimony blood test, and on November 18 we made our vows. There must have been close to two hundred people at the wedding, and boy, did we have fun! Lots of moonshine, and lots of dancing.

I knew I had to find a home for my new bride. She was only sixteen but she was quite capable of running a house. We decided that a move to B.C. with my family was a good plan. But before heading west, we spent a very brief time in snowy southern Saskatchewan. Along with my folks, Phyllis and I rented a little shack near Brooking, my birthplace. I found work skimming salt off of frozen lakes. But the weather was too harsh and within less than a week Phyllis and I returned to the empty homestead, (the only thing my folks hadn't sold) where I made ties for the rest of the winter.

I worked with a fellow named Neil Rockwell hauling ties non-stop for eight days straight. It was about a forty mile trip, back and forth. I'd sleep one way and Neil the other. It was handy that the homestead was on our route, so we could stop in for a good quick meal from Phyllis.

By the early spring, Mother and Dad had gone to the west coast and come July, a pregnant Phyllis and I were close behind. It didn't bother me that we were leaving Saskatchewan. I was twenty-two years old and was looking forward to a new place to call home.

10

A Move to B.C.

I have a crazy memory of my first driving experience in Vancouver. I was used to the flat land and open roads, so my biggest problem was trying to make my way around the street cars; there seemed to be tracks and cables everywhere. Phyllis and I stayed in the city with my cousin Ethel for a couple of days before we headed over to the Island. There were no B.C. ferries at that time. The CPR barge for railroad cars took vehicles to Ladysmith, while passengers traveled on a different boat. The fee was $75.00 for each truck to be transported over to Ladysmith, then you'd drive to your next stop. It was the middle of July and because of construction on the highway to Nanoose Bay it was a hell of a mess. It was a rough road, paved only as far as Campbell River, and it took us a full day to drive that narrow, crooked road before we finally reached the Comox Valley. Joe and Anne had already purchased a place in Black Creek. It seemed to make sense that Dad and I would purchase land there too.

I must admit I wasn't crazy about the Island at first. In fact, it took me about four or five years to finally feel at home here. It may have had something to do with my fear of water, but I also hated the feeling of being fenced in. And with only one road down the island, I felt restricted. I left Phyllis with Anne while I went to look for work. My first concern was finding a good home for Phyllis and our child on the way. I was always quite religious about being financially ahead.

They were just building the John Hart dam, so Neil Rockwell and I drove up to see if we could get a job. I was offered $3 an hour as a truckdriver. Considering I had made $75 a day in Saskatchewan, I declined the offer and headed up to Prince George to find work in the lumber business.

I loaded my truck onto a barge (which cost me another $150) and headed up to Prince George. It was disappointing to have not found work on the Island but I didn't really like the feeling of being "hemmed in" anyway. I worked in Prince George for a short time before the cold weather came and work petered out. By the first of November Neil had decided to stay in Prince George but I went back to the island.

In the fall of 1947 I made a bold move. I sold my trucks, and with the profit I put a down payment on a piece of land in Merville. I was 22 years old. The land cost me $4,750 and people said I was crazy, because at the time it seemed a hell of a lot of money. It was the only place on Howard Road that had electricity. Phyllis and I furnished the little house with just a chair and a chesterfield. Times were tough and I needed to find work. My land payments were only $156.25 every three months, but I was struggling to make them.

People had the tendency to piss me off, but I was never mad at money. In January of '48, I found work logging. I had a bit of an attitude, but I was an honest worker. I went power saw falling with Big Burnett Power Saws in Kelsey Bay. Those saws weighed about one hundred and sixty pounds each. Joe Pederson was the faller, and a fellow named Howard Ralston was the head faller. It was a new job site for me and that meant learning new rules. On the first morning of work I

went into the filing shed where I was given a saw to use.
Each worker had their own. I chucked my saw into the
carrying rack in the back of the crummy and we headed to
the site. When we got there I grabbed the first saw within
reach and turned to climb out of the truck. I didn't get far. A
big bucker tapped me on the shoulder and said, "Hey, that's
my saw."

He knew I was "green", and I didn't even remember where
the hell I'd put my own saw. I decided it would be best for me
to hang back and wait, knowing that the saw left behind would
be mine. After that, I tied a little piece of cedar around my saw
so that I wouldn't make the same mistake again.

I found faller's hours pretty easy compared to what I was
used to working. We worked six hours a day, five days a week.
And there were benefits, too. Joe and I felled the biggest cedar
up there at that time. It took us all the whole day to knock it
down, and when we did the stump measured 16' 2" across. We
had to apply side notches, wedges, and a lot of sawing, before
that ice cream cone would tip. It was massive.

We worked in the bush for three or four weeks straight,
until around February when we got a hell of a snow. The saw
contractor, Stormy Burnett (his brother invented the "Burnett
Saw") had an old Model A car which four of us piled into to
head home. But we should have left a day or two sooner. The
snow was coming down so hard that little Model A didn't
stand a chance without our help. The old road was steep, and
crooked as a dog's leg. It took us four days to get out, taking
turns shoveling the snow from Kelsey Bay to Campbell River.

When April rolled around, I quit and went to work for
"Van. West" up on Cumberland/Royston Road. I remained
with them for about a year before quitting to work for
Nanaimo Lakes. After staying there for only three days I found

work with Mahood Logging. There I remained until I got another mad-on, which in those days seemed to happen quite often. To me it was simple: I worked hard, and I expected an honest days pay for it. If that didn't happen, I'd be damned if I was going to stick around.

You had to watch the scalers closely and I had had altercations with them on more that one occasion. Like the time I was working up around Jarvis Inlet. We went up there in the Spring and worked for ninety days straight. I never missed a day. In those days a good faller would make about $20 a day. If you had an especially good day the company would hold a bit of money back from you, and if you had a bad day, they'd give you some back. Well I never liked that method. If I had a bad day and a small scale, I could deal with it. But if I had a good day, I wanted all of it. The boss was the scaler and this one day the head faller and I were suspicious of him cutting us short. I decided to confront him on the matter. He was saying, "Bud, I admit I'm tight, but I'm not a stealer."

I cut him off quick and said, "You're not tight. You're a fucking thief!"

I knew I was finished working there, but I couldn't leave right away. I had to sit across the table from the guy for three days before the boat could come and take me out. You can bet we didn't talk much.

I've worked with some real characters over the years. One who really stands out is Lou Clark. He had nine kids, and he used to perform this joke he thought was pretty funny. He'd line his kids up holding hands, with his dog on the end, then he'd grab a spark plug on his Model A car and you could see it go BRRRRRTT all along the line. The kids would laugh so hard and that poor dog would yelp. I guess in a way, Lou was like a big kid himself.

He had a thing for sparks. We were falling in quite a bit of snow one time and Lou was using his usual method of fixing a flooded saw. He'd take a spark plug and light a match to burn the excess gas out of the top of the piston. But that time the saw caught on fire, so he ripped his coat off and threw it over the saw to smother the flame. It didn't work completely, so when our boss came along Lou didn't think twice about grabbing his coat to smother the remaining flames. It burned up too. Eventually, Lou did get the fire out but he had to use a handsaw for the rest of the day. He also had to buy a couple of new coats.

Ollie Pederson was another man I worked with for a period of time. He was a solid worker but he liked to have fun too. Ollie had spent some time overseas and he never quite seemed the same once he'd returned. If he had too much to drink he'd think nothing of jumping through a window. One night Lou had a party and I guess Ollie had to go to the bathroom. He jumped up onto the chesterfield, dropped his drawers and did his thing right there in front of everyone.

But I think the funniest thing he did was after a party at his house on Coleman Road. His wife got up in the morning to light the stove, and almost immediately she could smell something bad. She thought something had crawled in there and died. She saw a chair sitting beside the stove, and put two and two together. She ran into the bedroom where Ollie was still sleeping and jerked the covers off. Sure enough, Ollie had a black ring around his bum. In his drunken stupor, he'd crapped in the stove. His wife woke him up quick and said, "Ollie, you gotta come and light the fire. I can't get it going."

It took him a while to drag his sorry ass out of bed but it didn't take him long to discover the problem.

I continued falling until around 1952 and I was making

good money. Regular day wages for fallers then was probably only about $13 a day. Ollie and I worked together a lot. He was an excellent partner.

When we worked for Carter and Elliot at Saratoga Beach there were some days when it was so windy it was too rough to fall. Ollie and I would sit in the crummy talking, waiting for the wind to die down.

Another time, we were falling out at the top of the hill up in Comox around the Indian reserve. The deal the natives had made with Carter and Elliot was that they wanted to fall the small trees and leave the big ones for us. The trouble was they started about a month before we did which was the wrong way to go. Falling those big trees over the little ones would have left quite a mess. As it was we had a hell of a time trying to wedge the big timbers to save the small stuff.

One day, just near quitting time, Ollie and I had cut through this big tree and it was just starting to tip. We saw these five Indians coming through the brush, and being the hyper guy he is, Ollie started screaming and hollering for them to get out of the way. He was sure they were in danger of being flattened. That night when we told our boss about it he just laughed and said, "Goddamnit, if you'd seen a puff of feathers, you'd known you got one."

Early Days

Me and Dad before
the road trip north

The day before our trek north

Nipawin - 1920's

Me on our car - 1928

Tractor and combine - 1929
owned by neighbour, August Labelle

Mary on our pony - 1922

Me and Poodle - 1937   Mom on Gypsy - 1930

Me and dog team - 1936

Uncle Burt's house

Threshing - 1940's

I fell my first tree with Bo Anderson - 1932

Dad, Mom, Anne and Mary with the
Russian Wolfhounds - 1920's

Snowsedan

Me shaving
*(gettin' prettied
up to go girling)*

1938 - Hauling logs
with a sloop

Cold Decking Logs - Bottings Mill, North Sask.

Me with two work horses

I sold fifteen head of cattle for the grey horse
while my parents were away

Me and Marion Conibear at Joe and Anne's wedding - 1945

Joe, Anne and Ollie - 1945

Dad, Mom and Grandma
Mary (at age 94) - 1954

Cousin Hilda and me

Me and Dad in front of the homestead

My first sawmill

Skunks and wolves
Dad caught

Me going hunting on Dainty - 1938

Me, Dad and Bill Smalley,
a drifter - 1937

The Homestead

Me, Phyllis and friends - 1945

Me - 1941

Later Days

Gary Ellis

Resting on the mountain range along the Fraser River

Anne's record-breaking Caribou. Its horn span measured eight inches wider than the recorded world record.

A horse's view of the Itcha Mountains

Packing things up on another caribou trip

CJC Equipment Sales

Chezicut Ranch Branding - 1983

Dwight

Dwight's 16th Birthday

Sharon

Sharon and Wayne

Mother at 84

Anne and I hunting

Three generations - Bud, Dwight and William

11

Home Again

I finally took a break from work to head back home. In January 1948, Phyllis had given birth to a baby girl. I was away working and was eager to see her. Sharon weighed less than five pounds and it was quite a moment to see her tiny body in the incubator. When I saw the knot in her navel, I said, "By God, she's a born cowgirl — she's already tying knots!" We bundled her up and put her into this little grape basket to take her home.

I spent a little more time around the home then and thought about clearing the bush around the property. Phyllis and I had saved enough money to buy the odd cow which we raised and peddled. It wasn't uncommon for me to butcher five cows after supper. It would take me about forty-five minutes to skin, gut, and split it in half.

For a while there I was selling about a ton of beef a week to a Cumberland butcher named Mr. Murphy. But he was only able to pay me about 45 cents a pound, so eventually I sold to Overwaitea. By then I was butchering up to twenty cows a week which left us with a good profit, and a nice supply of prairie oysters (calf testicles). Phyllis never liked to cook them; something about the smell. But I'll tell you, you've never eaten anything good until you've eaten prairie oysters.

One time I came home with two full buckets of them and boy did we have a feast. We'd boil them, to crack their tight skin, then peel it off before splitting them in half to throw into the frying pan. They're the finest textured meat of any animal.

Around that time, I was awaiting the end of fire season and found a job building a barn for UBC Farms. The owner, Barret Montfort, hired a bunch of loggers and paid them $1.50 an hour to shingle the roof. The roof was huge: about 200' long, with a real steep pitch. Two of us put those shingles on at a dollar a square. It took most of the summer to build the barn and the cows had a home by fall. When the rain came the other loggers went back to work in the bush, but I'd made the decision to stay at UBC Farms.

That fall, just after a carpenter had built a new box on the truck for the cattle, we were transporting some cattle to the MacAulay Road farm. Turning onto the bridge, just in front of the Fisherman's Lodge, the cattle shifted in the back. The side of the box busted open and they all spilled out. It must have been quite a sight to see. Cow chaos. We chased them around and finally got them back in, except for one little heifer. I grabbed my rope and ran like hell after her. I chased her around until my tongue was hanging out of my mouth and I finally roped her. One of the guys had to sit on her for the rest of the ride to make sure we got her home safely. It was quite an ordeal.

After working for about eight months with Island Ready Mix, Montfort offered me a job back at the farms as manager. I didn't hesitate. I quit with Island Ready Mix that night. Montfort came to me to offer a wage. He was pretty tight with his money then. I guess that's why he had so many millions. I was making $1.40 an hour with Island Ready Mix and Montfort said he could beat that. He wasn't in it to make money.

Managing UBC farms was quite a job. There were probably one hundred and fifty head of cattle on that thousand acres. In my ten months there, I calved twenty-nine cows, including the heifers — I had thirty calves on the ground. I'd learned cow birthing through experience. Once in a while, I'd have to deal with a calf coming out backwards and I'd have to flip it around fast. A lot of it had to do with being calm, and common sense. Everybody knows what a calf looks like — you just had to make sure it was coming out the right way. Of course there was the odd stillborn, but that was to be expected. Montfort was negotiating a deal with UBC to either give them the land or he'd sell it to them for a dollar. But he "assured me," part of the deal would be that I stayed on as manager.

I knew damn well that wasn't going to last too long. He spoke to me in November and UBC took it over on the first of January. From then on I had to answer to them. I had to phone them every Wednesday night to update them on things. Which was pretty tough, considering not much changed from week to week.

There was a fellow named Frank Netto who had worked on the farm from the start who caused me some problems. I guess he was telling Montfort that it was him who was calving all the cows and it wasn't long before something had to change. When the phone rang on Friday morning I had a feeling that something was up so I told Frank to come along with me to answer it. Sure enough, the fellow on the other end of the line said, "Well Bud, we've got different plans and we really don't need you anymore. I'm giving you six weeks notice."

"You've already given me the notice," I said. "Give me six weeks pay, and I'll be out of here by tomorrow night. Who do I give the key to?"

"Oh, Mr. Netto will be looking after things now," he said.

I wasn't surprised.

The next morning, Frank showed up on my doorstep in a panic. "God, Bud," he said, "there's a cow in trouble over there. She's calving!"

I said, "Jeez, Frank, I don't know nothin' about this. I thought you were the one doing all the work."

Before I closed the door on him, I said, "You've got to do one of two things — either get in there and jerk that calf out yourself, or phone the vet. I'm not coming back to the barn."

12

Dwight, My Son

He was born in 1953, weighed eight pounds, and he always seemed to be hungry. Dwight was never happy with baby formula. He was always squawking — he needed double strength. I used to give him a piece of pork rind to suck on. That'd keep him happy.

Dwight grew up fast but when he was eleven, he went into Victoria hospital for a bad infection which at the time they thought was leukemia. He remained there for three or four months right through his twelfth birthday. The doctor said he didn't know if Dwight was going to make it and that if he did he'd be in a wheelchair. I smoked a lot then, and drank pretty hard, too. But when Dwight had to spend his twelfth birthday in the hospital, I made a choice to quit smoking. I threw my last cigarette out the truck window and vowed to never smoke again. I thought in some silly way that if I gave something up, Dwight would get better.

He survived, barely, and was grossly overweight when we finally welcomed him back home. At over two hundred pounds, he had no cushion left in his vertebrate and was in a cast from his knees to his neck. He basically had to learn to walk again. It was so hard to see him like that.

But, as my son got older, the problem seemed to go away.

All he ever wanted to do was drive trucks and even though I knew it was the worst thing for his back, I also knew that it was his number one love. I gave him a cow truck on his sixteenth birthday and he never looked back. Within a year, that truck had over 100,000 miles on it.

Eventually, with three trucks and five trailers in his ownership, Dwight started up Sointula Transport. The tougher the going, the better he liked it. When he was twenty he married and eventually had two kids, Wes and Toni-Lynn. Through it all he kept on driving truck. In fact, I remember at his engagement party he said, "Just remember Yvonne, my first love is the truck."

I think the only thing my son and I argued over would be who got the last prairie oyster. We'd have eaten a dozen each before they even had a chance to hit the frying pan. One time we took a bunch down to the neighbour's and they commented on how they'd had a lot of mountain oysters but none as good as ours. Dwight said proudly, "That's because they're prairie oysters." That's just what we called them.

Dwight and Yvonne divorced and years later, he met Laurie with whom he had two boys, Darwin and Justin. He and Laurie never did marry. The wedding was all lined up, and Dwight said to me, "God, Dad, I don't want to get married." He phoned her and said it was off.

Things took a real turn for the worse. Dwight moved to Vancouver, and on June 13th, 1982, he died. I was told he'd been at a friend's place for a barbecue and had said 'Goodnight' at 11 p.m., before he climbed in his truck to go to sleep. The next morning, I got a phone call saying Dwight had burned to death from a lit cigarette in his truck.

Well, I went ape shit. I just lost it, and was drinking pretty heavily for a while. To this day, I still don't believe he burned from a cigarette. He never smoked in his truck and when he crawled into the sleeper at night, he always left a window open for ventilation. I just can't imagine him going to bed with everything shut tight. And although I can't think of any enemies he might have had, I suspect foul play to this day. My son was a pretty popular guy, aside from the odd jealous husband. Years later, when my sister Mary died, mom said, "God, Bud, it's hard to bury your own kid." But it had happened to me. I understood what she meant.

We were really close, Dwight and I. I don't think there was a week that went by that he didn't phone me. I really miss him. I go to his grave next to my mother and dad as much as I can, just to say a little prayer and talk to him.

All my life I drank, but in moderation. I mean, I entered my first beer parlor when I was fourteen. But like I said, when Dwight died, I lost it. I drank like a sucker for a year and a half. You think you can drown your sorrows, but in my mind, it's a coward's way out. It doesn't make it better — it only makes it worse. For yourself, and for everyone around you.

Even when I was a controlled drinker, I never let a hangover slow me down. I drank in my truck while I drove, but I never let booze affect my work. See, I was always on the road, so I never had any time to stop and drink. A mickey (thirteen ounce) and a twenty-six ounce of rye, and I'd be good to go. I always called it "a cow and a calf." I'd finish those in the five or six hour drive between Kamloops and the ferry.

I did get picked up by the cops one time. I'd dropped a couple of loads of cows off in Langley and in Kamloops, and on my way through Qualicum I got pulled over. I got out of the truck which was carrying about four or five cattle, and I just handed the keys over to the officer. I gave him directions to Arrowsmith Farm where we were headed and said, "They'll look after the cattle, and I guess you'll look after me."

The cop drove me to the farm and dropped me off. The farm loaned me a hired hand from the farm, because I still had time to catch the last ferry.

I quit drinking in 1983 and I haven't drank since.

13

A Trucker's Life for Me

Ibegan my trucking career in around 1954. Over the years I've probably driven about thirty trucks, from logging trucks to cattle trucks. My first job trucking was with Island Ready-Mix. I drove for them for a couple of months before they took me off truck driving to set up a plant in Campbell River. I drove for Cliff Coulson for a couple of years. We did a lot of work in the upper Campbell River dam and moved houses out to Bevan Lodge.

I really enjoyed that job. There was this one trip we did in the spring that I remember particularly well. We had to haul three D8 Cats up to Faulkland and because of bad weather the roads were really bad, so there were some restrictions on. The other driver, Stan, got his load to the scales, unloaded, and went back home. The third transporter and I took our loads to the proper destination, then I came back and grabbed Stan's cat. It made sense to me to just finish the job while we were there. It was my first trip in the dark over the Hope Princeton Highway and it was a hairy ride. I was driving an old K-11 International truck, and suddenly the sign came into view: "STOP AND CHECK YOUR BRAKES." I put my foot down hard but not a lot was happening. I couldn't stop. I started down Whipsaw Hill in the lowest gear I could manage and if it ever sounded like I was going to blow the motor, I'd slip it up another gear. I used every inch of that road to get myself safely to the bottom of the hill and I was whistling pretty fast. I made it safely but when I came back the next day, I realized what I'd come down in that frosty condition. I hadn't felt scared until then.

That same year I started hauling logs. I was a pretty wild driver but I only had one close call. I met a load coming down the hill and the other driver didn't give me as much room as I needed. I hit the ditch, went down the bank, and the logging trailer flipped right over and landed in front of me in a swamp about eight feet deep. The boss came along shortly after and all he said was, "What the hell are you doing, Bud? Looking for frogs?"

Somehow, the truck remained intact.

My trucking career ended when I had a run-in with a guy we called "Intravenous". His real name was Venus, and he was around when I broke a stake-line and lost a load of logs on a dusty road. Venus was following too close behind me to avoid them and he wound up straddling a boom stick, leaving all four of his wheels off the ground. I can still remember the look on his face — he was so mad. I just stood there laughing. Venus called the boss, Cliffe, that night, and told him, "You either get Bud off of the road, or I have to fire the truck. There's not enough room on this road for both of us, and I was here first."

I knew something was up on Saturday when I went to work on the trucks with Cliffe. He was quiet all day. We stopped in at the Fisherman's Pub on the way home for a beer and the next evening, he finally phoned and said, "Venus doesn't want you on the logging road anymore."

I was disappointed, but by that time, I'd also been busy

with cow trading so it didn't matter much that that was the end of my logging truck driving career. I turned in my air driving ticket and continued driving low bed.

14

Cow Trading

Cow trading is to this day, my favorite career. I really enjoy buying and selling. Partly because of the profit at the end, and of course, because it's a challenge. You have to be able to outwit other dealers, have a home or a market for what you're buying, and of course, to be able to sell at a profit.

During the time I started seriously cow dealing, I had a home for every type of cow — feeders, steers, heifers, bulls or cows. I always had orders for every kind. Once the bidding began, I was pretty hard to beat because quite often they'd want half of them but couldn't find a home for the others. Some of the cows went directly to the packing house, some went straight to the feed lots, and the others I'd bring home. It's likely that having a home for each of the cattle was a key to my success in the cow trading business.

When I first started I bought this old '47 truck for $700 and I had a load of cattle of my own that I'd raised. The local dealers told me how little those cattle were worth so I'd go to the packing house myself to make sure that was the case. I was determined to make sure that if indeed they were worth nothing, I was there to witness it.

The very first calf I raised I bought from a neighbour for

$2. Back then veal was worth 30 cents a pound or less and by the time I was ready to sell the calf was up to two hundred and fifty pounds. I put a price of $40 on it. I'd gone fishing the day the prospective buyer came around. He was at the barn getting ready to load up the calf when Phyllis approached him. She was making sure he had the money. He said to her, "It's not worth $40. Just tell Bud to stop in on his way home tomorrow and he can see its weight." But Phyllis wasn't budging. "You'd better put it back in the barn," she said, "because he wants $40."

The man left frustrated, with an empty load. The next day a fellow named Jimmy Mitchell, who used to peddle fresh meat from his van, came around and he needed some veal. I showed him the calf and said that I wanted $40 for it. He said, "Oh God, Bud, it's worth more than that. I'll come and get it tonight." He gave me $47 for it.

Investing my time and money into cow trading is what got me where I am today. And for a while there, when calves were cheap, at about 12 cents a pound, I was providing most of the local farms with cattle. I'd frequent auctions in the interior mostly and was pretty busy at it every year until everybody else started doing it. Eventually, it just petered out. When I started doing it there were hardly any beef cattle in the Valley. The rest, as they say, is history.

Back then, cattle dealers were pretty cheap. If I bid on a cow, chances were good it would end up on the back of my truck. Cow buyers had a bad reputation. Quite often they were chased out of people's yards. They were thought of as crooks. I never was chased out of a yard. And there are a couple of rules I've established over the years: don't phone any cattle dealer after 6 a.m. and expect to find them at home, and, let the rancher put a price on it. If you like the price, buy it. If not, don't.

Usually, I'd put the money on the line. If they didn't accept it then, they'd phone me and tell me to pick it up the next week because the competition wouldn't have beat it. As the market dropped I could always depend on what I got for last week's load.

Jimmy Baird was the first person to start a feed lot, so I marketed through the company he bought for Canada Packers in Richmond in the beginning. But in the early sixties they shut down and moved to Alberta.

I didn't exactly keep track of all my business then and the tax department caught on pretty quick. Walter P. Duck was the first guy to come by for an audit, and I had no receipts to show. He said, "You sold $21,000 worth to Canada Packers, and you have no receipts to show for it. We want seventeen percent of your profits."

Not if I could help it. Over the next six months or so, I went around to every person I could think of that I'd ever bought a calf from and got a receipt for it. By the time the smoke cleared I owed the tax department $3.75. I later received a letter from the government saying, "No further comment for the time being." I never did pay them. I knew that it wouldn't be the last time I'd have to deal with the government.

I've made some great friends over the years, and one of them is Bud Noel. We first met at a Fraser Valley auction and from then on we partnered up regularly buying cattle and horses in the interior. We've got enough stories of our business ventures together to fill a book alone. One time we bought pretty close to one hundred Indian cattle out of the Fraser Canyon. We had to cowboy them out for two days up near Lillooet and we lost two along the way. They were wild ones. We got the cows down to Earl's Court, a ranch in Lytton, where we kept the cattle until we could load them on the cable ferry at Lytton. You could only put five cows on at a time, and with one hundred cows to transport it took us more than a day to finish the job. But for a case of beer, the ferry guys would usually give you an extra ride.

There were days when you'd ride around all day on the native land and you wouldn't see a cow or a horse, even though the natives said they were there. On one of these days, it was getting on towards the evening, finally I said, "I gotta go." I had a sense of where my truck was parked miles away over the mountain. I told the Indian, "I'll just leave the saddle laying there and turn the horse loose." And that's just what I did.

I used to buy a lot of wild horses from that area. One of the natives, named Percy, was just a hell of a guy and a pretty good cowboy, too. I'd always break out a gallon jug of wine when I'd finished dealing with the Indians. I'd have one little swig with them, then I'd go. I knew that next week they'd be phoning with another load of horses around.

By then it was 1953 and Black Ball Ferries was the transport system across the Georgia Straight. That was before Premier Bennett got in and renamed it B.C. Ferries. With all of my business dealings going on taking the ferry was a regular thing for me. Ferry fees were $12 one way for a truck, and $5 for a car and driver. Gas was cheap at 40 cents a gallon, and at a $10 profit per cow, you get nine cows in a truck. That had $1.40 an hour wage beat all to hell. And, I was doing what I wanted to do.

The first ferry to do the Vancouver Island run was called the Chinook. It was pretty odd to see a foot passenger on board and damn near as odd to see a car. State rooms were available to passengers and the cafeteria was more like a restaurant.

Waiters would come to your table to take your order, which for me was usually a good thick steak and pie for dessert. They really catered to the truckers and the food was excellent.

At the end of the meal the deck of cards would come out. The first ferry on Monday morning was the best because all of the truckers, especially the ones that liked to gamble, would be on it. Half of the crew would play poker with us. And even though there generally was a $2 at a time betting limit, it wasn't hard to lose a couple hundred bucks. It was really an ideal gambling location because we all knew that whether you were ahead of the game or broke, you had to quit when the ferry reached its destination.

Truckers were billed for ferry use once a month, and with twelve sailings a month, you'd get your thirty or forty percent discount. It cost a maximum of $90 for a truck and low-bed and sometimes, for a shot of whiskey, I wouldn't have to pay at all. Nowadays, they'd be afraid to give you a free coffee in case one of their crew squealed on them.

In the fall of 1953, Phyllis fell really ill. We know now that it was related to a nerve disorder, but at the time no one had really seen anything like it. The skin on her hands and feet was peeling off in clumps. It was awful. You could slide silver dollars under the skin it was so bad. We didn't have the faintest idea what it was. The Comox hospital kept her hands and feet wrapped up all the time and she damn near went out of her mind with the itch. Understandably, the least little thing would set her off.

She received care in the Comox hospital for about six weeks and finally went to see a specialist in Victoria where she stayed for six or eight weeks. It got to the point where I didn't see her improving; I decided to carry her out of there and get her home.

It was decided that seeing a chiropractor in Campbell River was the next best option. Within about five or six visits, he straightened Phyllis out.

Over the three or four months that this went on I juggled my work while caring for Sharon and Dwight: transporting them to my parent's home late at night and picking them up at three or four in the morning just so I could continue working. There was many a night when the kids didn't even know they'd been moved. I never missed a night.

I wasn't around for the kids as much as I'd have liked to be, with work keeping me on the road, but when Phyllis was sick I made sure Sharon and Dwight woke up at home. If I dropped them off at my folks in the evening I'd make sure I'd be back to transport them home before dawn. I always told them I'd pick them up later that night. And, although sometimes it would be midnight or one in the morning before I'd finish my haul, I'd take them home while they were sound asleep. I always had them in their own beds by the time they awoke in the morning.

There was this one time when I'd bought some cows over on Hornby Island and the waters were so rough that Albert Savoie, who owned the ferry then told me, "Bud, I can't cross tonight with you and your load, but I can take you over empty." The Denman ferry only had a six car capacity back then and the Hornby ferry could hardly be called that, with space for only one car. I kept my word with the kids and crossed without the load. Then, I went back the next day to pick up the load of cattle. That was the closest I ever came to not coming home for the kids. I kept my word.

15

Wild Horses

With Phyllis all better and spring rolling around I focused my time on buying and catching wild horses. We chased them in the Chilcotin, Fraser Canyon, and quite a few around Spence's Bridge. Studs would run off with a few mares and they'd just multiply. There were literally thousands of them roaming the fields before my time. A bounty hunter shot seven hundred of the wild horses one spring. Ranchers wanted them off the range, and forestry would give them $7 for their ears — merely proof that the job had been done. Douglas Lake Ranch used to use three hundred teams come haying season — that's six hundred horses, plus a hell of a lot of spares. And that didn't even count saddle stock. When tractors started coming in, the horses were no longer needed. We'd get six or seven cents a pound for the meat. The bulk of the feed went to mink ranchers and the other part went to humans. They were graded just like cattle. White horses couldn't be used for human consumption as they were usually considered cancerous.

It was normal for me to only get about three or four hours of sleep a night, and some nights there'd be no sleep at all. My body got to the point where five or six minutes' sleep would feel like a whole night's.

We did a lot of horse chasing with the Indians because they always knew where to find them. We bought some good broken horses from them too, until there weren't so many broken horses left. My friend "Chief" would call me up and say, "Bud, I got you another load of hawses." That was how he pronounced it.

Over the years, Bud Noel and I have bought hundreds of horses together. This one time, I'd made a quick decision to head up to another reserve around Douglas Lake to buy some horses. Bud said, "I might go up there, too."

I took off and on my way up there, I decided to stay in Merritt overnight. I'd made a plan to get to where I was going by 7 a.m. I didn't feel the need to get there too early. Driving out to the reserve, which was about fifteen miles away, who do I see coming in the other direction but Bud Noel. He'd already bought my load of horses and was on his way home, and he was laughing like a bastard at me. It was a little game we'd play with each other — but I got him back the week after that.

Transporting that many horses sometimes meant a lot of trips with the trucks. At one point, Bud and I had gone to Alberta and bought 85 horses in one shot. We trucked them to one place in Alberta and then shipped them home in railroad cars; some loads we shipped right into Courtenay.

There was this one time when we railed a load of sixteen Shetland pony mares that had been bred to a jack (stud) donkey, and man they were funny looking things. They'd all been packed into a box-car and were on their way up Island when I received a worried phone call from one of the train workers. He said that there was one foal who was sick and they didn't know what to do about it. I told him Bud and I would meet the train in Dashwood, which was on the north side of Qualicum. The train pulled up and I backed my truck up to it. When I poked my head inside I saw that the train crew had put 4' x 4' s all around that little colt at the back of the car, separated from its mother who stood at the front. That poor little thing was sucking away at the wood — anything it could get its teeth into because it was so hungry. The workers were scared to death that it was going to die.

I unloaded the ponies and got the colt in with its mother, then drove them up to Courtenay. I had a big horse sale I was trying to get them home for. We sold all of them for about $300 a piece. You could hold the mule colts in your lap they were so small, and the ladies liked that.

I've had some pretty hairy situations where I packed the truck so full of livestock it wasn't exactly safe to drive. In around 1954, I bought a load of cattle from Jack Johnson in Kelsey Bay. He chased the fourteen head into the back of my truck while I recorded them on the back of a cigarette package. We had used every inch of my truck and there were two calves to go. We dragged them up high on a hay stack and threw them in on top of the others where they eventually wiggled their way in. Well, then I had to get them all home. With so much weight in the back, more than the front of the truck, I was forced to turn the truck around in the middle of Robert's Lake Road and back up the hill. The night was so black, the beam from my flashlight was like a white bean in a black cat's ass. But I made it up the hill, got home at 3 a.m., and unloaded the whole load leaving the calves with their mothers.

Once the cattle were corralled, I reloaded a portion of them and raced off to catch the six o'clock ferry to get them to Canada Packers. A trip like that was not a bit uncommon for me. I'd get an hour of sleep on the ferry and that would be enough to keep me going for days.

16

Trapping

Throughout my whole life I've trapped animals. As a young boy, I learned from watching those who knew all the methods of trapping and skinning and I've carried those methods through to this day.

I find Conaber traps work the best. They're high quality in that they don't allow the animal to suffer too long. A quick death and no chance of escaping and walking around with a limb missing. I do use the occasional leg-hold set, but you lose too many on those. The quicker the death, the better.

Times may have changed but animals' characteristics haven't. Otters are still the hobos. They never really have a permanent home. They wander up creeks and mess around on the beaches. They're the wolves of the water. 'Coons will try to outsmart you, but wolves are the toughest to trap. Island wolves' pelts aren't worth much so I usually head out to the Chilcotin Valley in the interior and stay with my friends Bill and Fay Coulthard at the Chezicut Ranch. Usually I'll head up there early in the fall and catch up to fifty beaver in a week.

Of course, there is a different trap for every kind of animal. From mink, otter, beaver, marten, 'coons and wolves. I still own about three hundred traps all together. Last year me and my buddy Phil Birch, an avid young trapper, put out 140 marten traps on my line. The next day we found fifty-six marten which is a damn good record. Phil left to go fishing and I went back two days later and pulled 'em all. I had another forty. That was enough marten to last a while.

Trapping has always been a means of survival for me, but it's also become a bit of a hobby. In 1986 I took a trapper's course and purchased a license, which, at that time, was $125, plus an annual $17 renewal fee. There were ten or twelve others taking the three day course and I was amused that the instructor usually directed the others to watch and learn my techniques as I was the most experienced.

In 1987 I bought an auctioned-off trap line from the government. The auction was down in Nanaimo and the rule was you had to open the bid at $500. From there, the price went up at no less than $100 a jump. They had six or seven different trap lines to auction off that day, and I partnered up with Val and Gordy Lavoie and Frank Naswell. When the officials in the game department came to the trap line that we wanted, a lady from Gold River opened the bid at $500. A log scaler from Tahsis put a bid up for $600. I said, "Seven hundred."

She said, "Eight!"

The scaler said, "Nine!"

I shouted, "Two thousand!"

Well, the whole thing stopped, and the officials said, "Jeez Bud, I don't know if you can do that. Is that legal?"

I said, "Well what the hell's the difference? A bid's a bid. Why mess around a dollar at a time when you're gonna go there anyhow?"

So often that'll just spook the competition right out — and it worked that time too. Then that trap line remains yours forever, unless you miss a couple of years, and lose it.

Trapping season begins in the middle of November and goes until mid February. As soon as you start catching the odd female on your line it's time to pull your traps. You don't want to upset the balance by spreading them too thin. You want to leave your breeding stock.

It takes about a week to hang the skins and dry them before you can take them off the boards. Then they're hung in a cool place to dry more before you ship them off to a fur auction. There are about three major fur auctions in Canada. A couple in Ontario, and I sell most of my furs to Vancouver's Pappas Furs. Quite often I'll get a call from the Games Department to come and catch a nuisance for them. Other than that, I stick to my line.

I've had my share of problems. Protesters taking my traps, and even police officers who thought I was trespassing. I had a run-in with the cops when the owners of the Comox Airbase called on me to catch some beaver down by their river beside the Griffin Pub. The beavers flood the creek and then it backs up. And in a lot of cases if you pick the animals up and move them, they don't come back but other beavers will just move in. The gate man and I spoke, and he said that he would leave the gate unlocked for me so I could get in the next morning. There I was on the second day doing my thing when this police patrol guy came along. Sirens blowing, lights flashing, and his gun out. He was a little French guy who could barely speak English. Soon enough, there was another patrol car along behind him. He wanted to know what I was doing in there. I told him, and he said, "Let me see your I.D."

Well, I was in a hell of a spot — bushy, and thick with crab apple and willow. I said, "That's fine, but you can come to me."

I had a little snicker to myself as I watched him wade

through the bushes and then through the water. A few months later, when the Pub called on me again to help them with the river pests, I sent my buddy, Phil, to do the job. He had no troubles.

Every two years, preparation for the new kits takes place in the beaver's nest. The two year olds will clean the house of its old grass and sticks, and replace it with fresh stuff before the next litter. They will then get a little bite from the mother beaver; their signal to move out on their own. Up or down the creek they'll go, to start a new colony of their own. Meanwhile, the yearlings still remaining in the nest assist the mother beaver in keeping an eye on the new kits; generally four to a litter.

The beaver fur here is just as good as anywhere. In fact, it's better than that of the beaver up north. Swimming around under the ice, they tend to lose their guard hair. We get just as much money for the fur here as we do for the fur on the interior beavers. The pelts are of value but the tails get thrown away. There are tanneries that take them but generally they're not good for much. Mind you, there are some who find them a tasty treat. It's known as a delicacy to the natives; that and moose's nose.

I've had one of my trailers parked up on the river at the fish hatchery since about 1988 which I stay in while I check my line. Nowadays, I find I have to hide my traps well to keep them out of view of the environmentalists. I hide them so well even the animals can hardly find them. There are lots of bears along that river and I've shot some that were close enough to grab. Pretty near every night they'd jump up at the side of my trailer and rock the hell out of it, peeking in the windows.

About two years ago a bear started living right in my trailer. He crawled into my bed — wrecked it. He made it the way he wanted it. Another trapper came around and noticed the door kinda looking silly, so he poked his head inside and noticed this bear sleeping in my bed. He slapped the trailer wall and the bear jumped up and took off. About two years later, that same bear got into the trailer just before I started trapping. The bottom of the trailer door is broken and is open about an inch or so and one night that old bear got his claws in and was reaching around. I snuck up, turned the knob, and gave the door a swift push and hit the bear in the face. He jumped back and was gone.

At one point, that bear got inside my trailer, got up on the table bench and pissed all over the seats. God, that was an awful smell. It took a lot of bleach to rid the trailer of the stench. That bear came around a lot before I had to shoot it.

You used to be allowed to trap two bear per trapping line, now you get it as a regulation on your license. There are so many bears, they're overpopulated (they're attracted to the hatchery) and there were some days when I could stick my head out of the trailer and count ten of them cruising around.

Bears are attracted to the smell of marten bait — beaver. It makes excellent bait for anything; wolves, you name it. And you can eat it yourself if you need to, unless you get a young one, which can taste a bit like bark.

Bear meat is tasty if you get it when they're eating berries. But when they get around fish, it makes them stink so bad you can barely stand to skin them. Bear grease is excellent for many things: cooking, or oiling your shoes.

Trapping in the winter is a bit of a bugger. There were nights when I'd go to sleep with my wet clothes on because it was the only way they'd dry out. And there were crazy roads I'd have to travel to get to my trap line. Up north around Tahsis Road, I'd have to chain up all four wheels on the truck because the road was so bad with snow. It's mountain country and the snow comes down fast. It could be real nice at the bottom of the hill and by the time I'd get to the top, there could be six or seven inches on the ground because the elevation changes so quick. I'd get in there early, get the traps and get out. If the snow got too deep, I couldn't even find them. Those steep hills would be crooked as hell, and lots of ice. The water runs down the hill in the day time and freezes at night. The forestry used to maintain that road and they had a couple of graders on it continually. When it was snowing they'd be out there twenty-four hours concentrating on one hill each. They'd never catch up.

I've had a few hairy incidents on those hills. I remember one time, there was no snow at the bottom when I started up but I got less than half way up, spun out and started sliding backwards like a bob sled. My first instinct was to put my foot on the brake and pretty soon I was sliding out of control. I was heading toward the bank, so I let'er free wheel and steered the truck into the bank backwards. It was kind of sitting cockeyed and I finally got the hell out of there.

Dwight had trucked up there years before I had the trap line and he was always telling me about those roads. They were worse than I'd imagined and Dwight drove his rig over them. He'd come home and chain up all eighteen wheels. He loved it. The tougher the road, the better he liked it. And there were times he'd go over with two trailers and seven axles behind him. For the real bad hills, he'd take one trailer off and take them up one at a time.

At the end of trapping season, I bring the traps home and boil them in a huge tub with a pound of wax. Boiling rids them of any smell, and the wax keeps the metal from rusting up too much in this west coast weather. It also blackens the metal to add an element of camouflage.

17

The Comox Valley Riding Club and Those Rodeo Days

In the early sixties I became a member of the local "Exhibition Board". The government owned a section of land on Headquarter's Road, across from Vanier School, and one spring they put it up for sale. I suggested at one of our board meetings that we put in a bid on the twenty acres, and turn it into a rodeo ground. I figured they'd take any price for it so we put our bid for just under $5000. (Something foolish, like $4,986). Some members doubted it would go over, considering we had no real funds to back it up. I said, "If we sign a petition for it, they're bound to consider it, no matter how small our payments are. Even if it takes thirty years to pay it off. You'll never get another piece of property like that."

Well by God, after about three more meetings, it was agreed by all to put the bid in. We got it, and by the end of its successful first year, the fair had made enough profit to pay that bill off. The government also gave us a grant of about $2,500 each year to support us and initially that money wasn't being used wisely. With the portable buildings (army tents) we were using, that money was just getting pissed away. So at one of the meetings I volunteered to go to the Campbell River pulp mill. I knew the guys up there and I figured they'd supply us with a small donation of lumber. With two years of grant money, we'd have enough to build a barn. That's the way it went and that first barn still stands there to this day.

Once things were rolling with the rodeo, I didn't stick around. I like getting things started, but once the challenge is gone, so am I. In 1961, a group of us pooled our resources to start the Valley's first riding club. There were ten or twelve of us who were really active in getting it started and we held it at the fairgrounds on Headquarters Road. Bill McLeod was our first president and many others helped contribute to its success. There was Bob and Marg Owens, from Victoria, Hunter Babcock, Jeff Latrelle, the Lovell family, and the Kvisle's.

It started with a small meeting at one of the member's houses and before too long there were well over one hundred members and more on the waiting list. We had to start screening people because we wanted it to be a family oriented club. No rowdies allowed.

With no lack of positive energy and motivation, we built an arena for damn near nothing because most of it was donated: fence posts from Comox Lake, and lumber donated by the Campbell River Pulp Mill. Within a couple of months, evenings included, the arena was completed, and on the May 24 weekend we held the first Courtenay Rodeo. Close to six hundred people showed up that opening day, locals and people from the mainland, as well as Rusty Russell, of the Cloverdale Rodeo. It really felt great to be part of such a large community event.

Admission for spectators was $2, $10 entry fee if you were participating in the events. That money went into the 'purse'; the prize purse was $75 for the five major events, which was pretty big for such a small time rodeo. Events ranged from saddle bronc, bull riding, bareback, steer wrestling, and calf roping. And women's barrel racing, as well.

First prize was a belt buckle which cost us $12.50, and the first one went to Mel Highland for saddle bronc. He later went on to become the world champion. I'd let him take a little horse home to practice on, a horse we'd named "Kruschev" (after the Russian leader, Nikita Kruschev). Earl Underhill won for calf roping. Barry Warden won for bareback. Dwight, who was eighteen at the time, was real good at cow mugging; roping a cow, while the other guy quickly gets a squirt of milk in a bottle.

But soon we realized that "Cumberland Days" were held on the same weekend so we switched the rodeo to Labour Day weekend. At around that time we had also established a riding club just for kids. The 4H Calf Club consisted of three age groups which we drove around to every horse show on the Island. From Campbell River and Port Alberni, to Victoria and the Vancouver PNE, I hauled calves and 4H members around for twenty-six years. But I still made time to do my cow and horse trading on the side.

I was purchasing new horses at least once a week. Most of the horses the rodeo was started with were purchased from the Indians. This particular one was quite a story. I went to Merritt to buy a horse from my friend Smitty Bent, who I called Sarge. (He'd been a sergeant in the second world war, and he had the nicest handwriting of any man I ever saw.)

We had planned to meet on the railroad tracks early one morning. As I saw Sarge leading the horse towards me, I should have clued in right then that there was something wrong with that horse. I could tell by the way it was walking. As the chestnut mare neared, I could see the big saddle-scalls on her sides, like she'd had the life ridden out of her. I asked one of the other natives about her, curious as to what might have caused her to be so wild in the saddle. They said that about three or four years before, Sarge had tried to ride her and she'd gotten away with the saddle still on. Apparently it had stayed on her until it finally rotted off which would probably have taken eight months or so. Hence the saddle scalls, and her ability to kick so damn high every time a saddle came near. She never got over that.

I decided to buy her anyway and when I got home I phoned up the man I'd had in mind for her. He was buying it for his daughter, and said he'd be up within the hour. It occurred to me then that maybe I should throw a saddle on the ol' horse myself, just to see what she was like. Well she threw me as high as a house, and I reconsidered selling her. Instead we harnessed her great bucking skills and named her Miss Courtenay. For years she was a hit at the Courtenay Rodeo before we sold her to the Riding Club.

I bought a black mare at around the same time from a fellow named Willy Mike. I went to halter-break her but she was mean. Meanest horse I ever met. I named her Black Bess. Her head was 1½ sizes bigger than a normal horse, I mean, she could hardly lift it. She'd grab the rope with her teeth, just like a dog. I'd never seen anything like it. And when I put a saddle on her, she'd lay down. When I had her bareback, she'd stand up. Normally, it would take me about half an hour to halter break a horse. With Black Bess it took me all day. I just couldn't budge her. I finally won her confidence by having someone lead her around. Eventually, we did make a good rope horse out of her but come winter, we let her go back up to York Road where she came from. We never saw her again.

Another time I purchased five or six good bucking horses in one load from the Douglas Lake reserve. I took four of them over to the Cloverdale Rodeo to be tried out. I sold those horses to Joe Kelsey, a stock contractor, for $75 a piece and we put them in the wild horse race. But they were pretty snaky so we made bareback horses out of three of them and a saddle bronco out of one. Three of those horses made it into the national rodeo finals.

Over the years I've come across some really excellent bucking horses. One sent its rider level with the electrical wires and it was still tied up! That particular one was actually so gentle you could lead it with a string, and I decided to buy it. I put it in the truck, which was covered by a canvas canopy, and made my way down the Iron River logging road. I'd tied the horse up loosely, with a lariat around its neck, because it was so gentle. Well I guess he got spooked in the truck being by himself or something, and then he wasn't so gentle. He reared up, tearing through the canvas, and busted the 2' x 4' the lariat was attached to. Then he leaped up over the front of the truck and smashed out every window (except the back one), folded the door down and kicked the steering wheel. My buddy Bob Owens was with me and with my door bent right down from the hinges, I was trying to steer away from the ditch that lay ahead. I got off balance so bad that we finally tipped over. When I jumped out the hind wheel burned the hide off of my leg a bit, and I thought Bob was dead for sure because of the way the truck was laying in the ditch. I scrambled up and there he was, sitting wide eyed. "He went through there, Bud," he said pointing a shaky finger.

A few days later a group of us went back and for three days tried to catch that horse. He was smart and he knew that brush country well. We'd get sight of him and then he'd take off and hide in the thickets. We ended up driving about ten or twelve of Bob's dude horses up there and after a couple of days, we finally caught him and cowboyed all of the horses out together.

Back at the truck, I loaded that horse in the back and tied its tail to one end of the box and its neck to the other. I was planning to take him down to Mount Vernon to the rodeo there. I made it home and phoned the vet to come and give the horse a little shot of tranquilizer to settle him down because he was dripping with sweat. I needed him calm for the journey ahead. I didn't have any windows in the back of my truck, so when I heard a helluva "thud", just as I passed Union Bay, I stopped fast and went to take a look. That poor horse was stone dead. I don't know if he'd had a heart attack or what.

They say that with a fat horse, sometimes if they get too hot the grease will melt right into their bloodstream and cause a heart attack. Some people thought it was the tranquilizer that caused it but I dunno. I went right back into the country where we caught him, and laid him there to rest. I figured it was where he wanted to be.

In the mid sixties, I bought thirty-eight head of heavy horses and brought them down to Oregon. Bud Noel and I prettied them up and sold each one of them in the auction sale. There'd be lots of partying in those days, but never before we'd finished with the chores.

It was steak and eggs for breakfast, and another steak for dinner. There wasn't any time for lunch, but we'd throw in a bottle of whiskey around that time. Always room for whiskey.

And a couple years before that, I'd bought a team of work horses and sold them to a horse logger down in Errington. It was a good investment. When Bud and I got them over to the scales, they weighed in at over a ton a piece. It was a pretty exciting three or four days. We had lots of fun and, of course, we made some money, too.

18

CCC Auction

It was 1964 when I began building the CCC Auction Barn. At that time I wasn't all that sure if it was even going to be an auction barn or not. I really just wanted a place where I could load and unload cattle undercover and close to the road. I went down to the beach and gathered a bunch of cedar logs, some fir poles for rafters and studs, and found a few telephone poles that had broken off for the posts. I got a big load of timbers from the pulp mill for $90 which was a hell of a load, for a deal. Heavy 4' x 12's. And with some plywood from Vancouver (for $2 a sheet, already painted), the forty-four foot wide, one hundred and twelve foot long walls were sheathed.

I built the auction barn by the light of the moon some nights and didn't even bother with scaffolding. I'd hammer standing on the tail gate of the truck, and when it grew too high I'd crawl up on the roof. I'd never built such a thing before and without so much as a square or a level to guide me, it didn't look too bad once I'd finished.

I'm very thankful to those who helped me. Sharon, who was fourteen then, and Dwight, ten, peeled most of the poles after school. There was a long list of kids who would swing by after school to help out as well. The tallest of the crew was

Randy Hodgins. He was about fourteen then and he was strong. He'd lift those 32' long poles up to the center of the barn and because he was afraid of heights he'd somehow manage to do a balancing act on a barrel on top of the truck to avoid going up the ladder to the roof. Wayne Jackson, Lyle Quinn, and Randy's cousin, Barry were a great help as well. It took the best part of a year to build that auction barn but it would have taken longer if not for the help of those kids.

By 1965, I'd become the seventh person in the Comox Valley to try a livestock auction barn. The others had lasted a year or less. Hard to say why they didn't make it. Maybe they gave up too soon. You have to be prepared to operate at a loss for a while.

I didn't realize that even though the building stood sturdy, ready for business, the tough part was still ahead. I had to deal with regulations. I phoned Victoria one morning and said I wanted a license to have an auction sale. "That's fine," they said, "we'll send you up our regulations. Then you can submit us a blueprint of how you're going to build it."

"Shit," I said, "it's already built. I just want a license."

Apparently it couldn't be done that way so I hired someone to draw me up a blueprint and he did it right to scale. A real good job. I mailed it down, stating that I needed the license by the 25th of March for the auction I had planned. They condemned it. They said the print was on the wrong kind of paper. I had to pay an architect $75 to transfer it to the proper kind of paper, before I finally passed that phase. They then sent up three different inspectors who kind of laughed at what stood before their eyes. "Strong and sturdy", they said, but it still wasn't proper.

Before they would pass it, I had to put in twelve inch cement footings in the sale ring. The toilet and the office I'd built seemed fine but there were a few things that weren't. Because the ceiling slanted down on that side of the barn, one side was a ten foot ceiling and the other side was six and a half. I needed an eight foot ceiling in the vet's office. It just wasn't good enough.

I fixed that problem quick. I took the power saw and whipped a foot off around the sale ring and put in some cement blocks. Then I fixed the vet's office by digging a hole into the gravel on the one end, to make it eight feet, and poured some cement in. It worked for the inspectors. Only a slight problem occurred whenever it rained, there'd be a foot of water in there.

The last hoop I had to jump through was with the bathroom code. But it didn't hold me back much. On March 25th, the CCC Auction building was approved and I was in business.

To me, CCC stood for Colbow's Cow Camp. But it soon became known as Complete Control Center. The structure of the building's lean-to ceiling allowed the auctioneer, and the spectators, to see everything. And in thirty-three years of operation, we've never had a complaint. The inspectors would have let us know, that's for sure. I actually thought that the auction was something that would only be temporary, but my buddy Mickey advised me to get a yard license and an operators license. He said, "That's something you should always own." He was right. The auction is a success to this day.

It was pretty tough going to start with, and the hardest part was finding a permanent auctioneer. There have been a few auctioneers work their magic at the CCC. The well known Frank Shuester, the Gibsons, Jack Marsh, Tommy Boyles and D. Edmonson all spent some time above the ring. And then there was my good friend, Gary Ellis. He'd come around to the barn sometimes in the evenings and pretend it was a live auction. At the time he was only sixteen, working in the sale ring because he was too young to have an auctioneer's license. When one of the auctioneers cancelled at the last minute, I knew that the show had to go on. People were already lining up. I was preparing to stand up there myself and start the bidding, when I saw Gary and I shouted, "Hey Gary! Why don't you get up here!"

He was a real natural and he's been doing an excellent job ever since.

The very first thing we sold at the CCC was a pair of pigs, to John Marinus. The next year, we expanded the barn and added a scale among other additions. Where we could originally fit one hundred and fifty cattle, we can now jam in three hundred. There's also a cattle squeeze for castration, if needed.

We really had it together, Gary and I. It was all very organized, right down to the signals we'd established with each other, and every time my hand hit the wall, it was time for another bid. I don't think there were two people who read each other better than Gary Ellis and me and there were days when it was tricky keeping control of everything. During one of our annual spring machinery sales, it was raining like a son of a bitch. The yard was full of machinery, two or three hundred items we'd brought back from the mainland and the interior. We'd attempted auctioning inside, giving each person a piece of paper with the items listed. It wasn't working. We were forced to stand in the rain while Gary's wife, Wendy, managed to keep track of everything until the sale was over.

Fred Lede had a cow auction in Black Creek that same year and that was a real barn burner. Everyone bought a cow. Fred would introduce each cow individually, to let the crowd know what was wrong or right with it. A good cow went for $800, right up to $1,700. It was a hell of a sale.

Of course, it wasn't all work. By June, Gary was becoming known for his auctioneering talent and he could sing pretty well too. He started singing "Happy Birthday" to Mrs. Lede and before we knew it the crowd had joined in. It was a memorable moment.

There was always room for the odd little fishing trip with the boys. After this one big horse sale Bud Noel and I had arranged, we decided to take off onto the open waters. The next morning we cruised out near Kitty Coleman, jigging for cod. We got to betting $20 a fish, who could catch the most. My buddy, Bud, beat me by two, even if they were hardly big enough to eat. When we got back to shore, Bud's woman was just madder than a bastard. She'd wanted to get back to Langley that evening and Bud had ignored her wishes. She got yapping and Bud looked at me and said, "C'mon, let's go back out fishing." We did and we didn't return until the next morning. By then, she was wilder than hell. She met him at the door and conked him on the head with her boot.

I've always been a firm believer in giving your word as a contract, so when Gary and Wendy Ellis took over the CCC Auction in 1987, we pretty much shook hands over a bottle of whiskey. They still run it to this day and I get my slice of the commission when the sale is over. And I still contribute by paying for any material needed for maintenance on the building and for operational fees. Diane Teesdale runs the hamburger stand and I get free coffee and whatever I want to eat. It's been a family operation from day one. Phyllis was in the office and my daughter, Sharon, did clerking. Before her, my niece Nina Elgie, worked there for twenty-two years before she passed away from cancer in 1996.

We started out strictly in livestock. We don't encourage miscellaneous stuff yet, but we'd auction it off in the first twenty minutes if we got any ducks or chickens. Regular auctions happen the second and fourth Saturday of each month. Feeder sales happen in the fall and the first Saturday in April is our machinery sale. We'll also slip in a few farm sales here and there.

The CCC has gone so smoothly over the years. The only slight catastrophe I can think of was the time we had some of those wild McCauley Road cattle to be auctioned off. Mr. Schultz, a cattle buyer, was sitting way up in the bleachers and it's what led us to adding reinforcements to the posts. One of those wild critters hit the wall of the sale ring and just about knocked it all down. It was right at the tail end of the auction so there were only three of four people there, but Schultz must have sensed something was going on because he started climbing higher. He was lucky he got out of that spot he was sitting in before anything came down. It was a close call. Needless to say, I've since reinforced the structure of the building so events like that won't happen again.

Scads of people used to come around for the horse sales as well. If it was a nice day, we'd sell outside and the horses would perform a bit better. It wasn't uncommon to have over three hundred people there for that. We'd sell tack for a couple of hours, then horses. And there would be up to seventy horses in there sometimes. In that case, it'd be a seven or eight hour sale.

Not all auctions went our way. There was "The Day No One Knew How to Bid" which happened in Powell River. Gary, Dave Elgie and I brought about fourteen horses over and I was sure it was going to be another "barn burner". We didn't get one bid all day. There were so many people there looking and I had a good feeling, but nobody seemed to know what to do. I realized they needed someone to get the bidding started, so I threw two dollars onto one of our piles of horse blankets. A boy jumped down, grabbed the horse blankets along with my two dollars. We had to chuckle; that was the only sale all day.

The following is a letter from the B.C. Livestock Producers Co-operative Association:

B.C. LIVESTOCK PRODUCERS CO-OPERATIVE ASSOCIATION

Auction Markets Located at:

KAMLOOPS • WILLIAMS LAKE • OKANAGAN FALLS
R.R. #2, Site 11a, Comp. 2, Kamloops, BC V2C 2J3

Whenever Bud Colbow attended an auction sale of any kind for B.C. Livestock, it was a given that Bud would receive #99 buyer number.

Bud put lots of enthusiasm into any sale that we held, usually sitting on the tailgate of our sound truck and bidding with his eye and if he got a little mad at the people bidding against him he would bid with his arm outstretched and bid like crazy.

One sale I remember like yesterday was the sale for the Douglas Lake Cattle Co. of their antiques and collectors' items. We had piles of old used horse shoes, old hard, used horse collars, and then some just real junk.

Ol' 99 started the bidding at $25; we went to about $50 a few dollars at a time, then he spotted the rich Vancouverite who was bidding on the stuff.

He said, "Hold it Larry, what's the bid?"

I said, "$50".

Bud said, "I'll give you $250".

I said, "Really!"

The man then raised his bid to $260 and Bud raised the bid to $400. The Vancouver guy raised his bid to $410 and when I asked Bud if he wanted to raise his bid he said, "No Larry, this stuff is just all junk, I was just trying to help you out". Man, I thought the guy he had raised the bid on would kill him. He continued like this all day. He likely made Douglas Lake Cattle Co. $10,000 or more that day.

Bud enjoyed the sales and we enjoyed him. Every time we came to something that we couldn't identify he would tell the crowd what it was and what it was used for. He had a great imagination and I think a good part of the time he had no idea whatsoever what the item was but a good story kept the people interested.

Cattle sales were also a time that Bud enjoyed as he sat there and bid against the "Big Boys" up on the cattle in the ring. He would bid on one head or he would bid on a liner load of them. It sure helped keep the regular buyers in tune.

Bud hardly ever came to a sale that he didn't have his truck and a load of something to put in the sale. He was loaded both ways, bringing some to us and taking some home... what a guy!

Ol' 99, believe you me, enjoyed auctions, people and a good time. Bud, you helped us all and auctioneering for B.C. Livestock was a lot more fun when you were on the tailgate or sitting in the front row.

All the best,
Larry Jordan

19

More and More Cows

There were quite a few wild cattle over on Quadra Island. About fifteen years before I started going there, some people at Maple Bay had bought some cattle and turned them loose. My buddy, Bob Owens, who was an excellent rider, came with me and we decided that we would just take the bulls. We were there for about three days chasing these eleven mature bulls. It was pretty tough going. Once we had them trail-broke, it wasn't so bad. We chased them over a mountain and finally got them into the monster size corral which our other partner, Bill Foort, had built the week before.

He was an incredible guy, Bill. Usually he'd chase those cows on foot and somehow he seemed to keep up with us. He was just tireless.

We got a bit of a chute built, loaded them up into the truck, and I hauled them away to Canada Packers. It took about two hours corralling them into the packing house. Those bulls were wild and they were so goddamned mad at you that you had to really watch yourself, making your way around on the fence instead of on the ground.

Soon after we went back to Quadra Island again and got four more. The cabin we would sleep in was good shelter from the rain and wind, but it was infested with rats. Bob was scared to death of rats and it was real dark when we got in one windy night. I snuck a little stick in and laid it beside my sleeping bag; I was up to no good. I knew Bob was scared shitless of those rats. When he'd finally laid down to go to sleep, I took that stick and gave the floor a good "whack!" He jumped up and shouted, "What the hell are you trying to do?" I said, "A rat just ran over me and I was trying to hit it."

Bob grabbed his blanket and outside he went. He jumped into the back of the truck and laid down on top of the dried horse-shit and slept there. I couldn't stop laughing. He never did come back into the cabin.

For a while there, I traveled to Denman and Hornby Island every week, but the first trip over was probably the most memorable time. My neighbour, Colin Strackan and I grabbed a couple of bottles of rum, jumped in my truck and headed to the ferry. I left my truck on the Denman side and we made our way to Hornby to fish for cod from the row boat. We drifted by a few farms that I couldn't pass up the opportunity to buy some cows from. After stopping at about four or five places, I'd purchased more cows than I could fit in the back of my truck.

For the rest of the afternoon, Colin and I fished just south of the Hornby ferry landing until we had about three sacks full: about three hundred pounds of fish. And, with all of the cows I'd purchased, I had to hire a freight truck to help with the load. We rowed back over to Denman, so I could grab my truck, but as Colin headed back across the straight, I realized that the ferry had just left. We'd finished our whiskey and I wasn't feeling any pain. I ditched the truck, literally, and hollered at Col' to come back. Clambering into the row boat, I slipped and fell into the cold March water. Soaked, and suddenly sober, we searched for shelter. An old abandoned house caught our eye.

The walls were filled with cracks big enough to throw a cat through, but we could see it had a stove in it. We dragged ourselves inside, lit a fire, and I bellied up to the warmth to dry my wet clothes, while I warmed my innards with more rum. The next morning, I missed the first ferry and the cows returned home to Phyllis before I did. Needless to say, I wasn't very popular around the house for a while.

In the late sixties, I made a cow deal with Dairyland. I had bought herds of cattle from Ruby Lewis and Graham Mitchell in Powell River before, and on this one occasion Dairyland had to get a large herd of cattle out of there. The cattle belonged to a Mr. Vlugt and Dairyland had to get rid of him before they could ship any milk over there. Dave Jones and Edgar Smith asked me to go over and price the herd so that they could then buy Mr. Vlugt's license from him. I priced forty-two dairy cows and then I didn't hear from them for a while.

About six weeks later they phoned to see if I was still interested in buying the cows. I said, "Well, a hell of a lot can happen to a dairy herd in six weeks," but I decided to go for it anyway. I bought the lot for about $14,000, and Bud Noel and I transported them back to the Valley in three truck loads. Once we got the cows home, we had to milk them and clip their heads and bellies for the next day's auction, which was scheduled to start at 1 p.m. We were a little concerned when 12:30 p.m. came and there were only about two or three people in the ring. But by 1 p.m., there were enough people to begin the bidding. By the time the smoke cleared that night, we'd sold every single cow and Bud and I split the $5,400 profit. Not bad.

20

The Tack Shop

In 1968, with the Valley growing the way it was, horses were becoming a big thing. I brought the first saddle horses in and was instrumental in starting the riding club, so I knew there was becoming a real need for equipment. There wasn't a saddle or anything around here for people to buy in Courtenay. I'd go to Langley just about every week and fill the truck full of stuff that people around here would ask me to buy. This went on for a couple of years. Well, one day I was sitting around with Jack Graham, one of the many cow traders I would meet for dinner at the Langley Hotel. He said, "Bud, why don't we go partners on a tack shop?"

Always up for a new challenge, I agreed. I came home and rented one of the buildings on Puntledge Road. I didn't really have any idea what I was doing, but as with everything in my life, I was prepared to learn through experience. I built shelves and stands for stock all day Friday, right into the evening. Saturday I went to the mainland and bought $36,000 worth of inventory (I could only pay them for half of it). With the new shelves well stocked with about one hundred saddles, as well as shoes, boots, clothing and feed, we were open for business Monday morning. We called it The Stampede Tack Shop. And boy, business went well. We were turning over a lot of money.

On Fridays we'd stay open until 8 or 9 p.m. We ran a real strict business financially. With Phyllis in charge and an invaluable clerk named Dolly Latrell, (she was a good friend of my daughter Sharon) they ran a tight ship. On holidays, when all other stores would close, I'd stay open. With a bottle of whiskey behind the counter, I'd have one hell of a day of business.

On an average, it was a poor day if I didn't sell a thousand dollars worth of stuff at the tack shop. We always made a profit, even during winter. And every month was better than the previous month. I must have owned it for about four years before I finally sold it. It was all new to me, but transporting the goods from the mainland was old hat. I was always over there once or twice a week picking up livestock and I'd never bring the truck home empty. More than half the time I'd be unloading the truck at two or three in the morning, so I'd be ready to go at six with something else. I never needed sleep. If there was something I could think of doing, it'd get done. Never leave for tomorrow what you can do today.

After six months of renting that building, I bought the lot just across the street where the Ford dealership now stands. I'd planned to re-build the store there because it would have better exposure to the road. But when it dawned on me that for proper exposure the store front would have to face the south east and you couldn't see any signs from that angle, I sold that property and bought and built the building that now stands as a furniture store on Puntledge Road. We added a ramp for loading up feed and expanded our space as we needed. Half was tack shop and half was feed.

In fact, we made some changes from the common use of hemp twine by introducing plastic twine to the Valley. We became the biggest retail twine dealer in the province. I'd order fifteen hundred bales of that twine (two balls of twine equalled one bale) and get it home early in the season. I'd haul it home via Parksville and Port Alberni and have two or three hundred bales sold before I even got home. I'd easily go through one thousand or fifteen hundred bales a year, right up until three years ago.

In 1972, I sold the business to Larry Holme for $135,000. Eventually, Bill Huxham bought it and changed the business name to Bavco. Me, I was moving on.

21

Yet Another Business

B y 1975, hobby farms were becoming popular in the Comox Valley. Business men were buying little acreages and I saw it as another profitable opportunity. I never missed a sale in the Fraser Valley, buying livestock and equipment. It wasn't a bit uncommon for me to drive away with two or three trailer truck loads and have half of it pre-sold before I even got home. If the price is right, you can sell anything. And with the prices of new equipment getting higher, it made it much easier to sell used stuff.

Sometimes Dwight would truck in a trailer load or two of equipment into Coombs and we'd pretty damn near sell it all. I didn't take any profit on those deals. I did it for fun.

I must have had a bit of spare time on my hands, because on the side I created yet another business: CJC Equipment Limited. I partnered up with Wayne Jackson and my son Dwight: Colbow, Jackson, Colbow. Our partnership was simple. I told them I'd work for $12.50 a day, seven days a week, just so that I could write the checks. That way I could purchase more items and line them up with more agents, etc.

But before we even had the company registered, Dwight came to me and said, "Dad, I don't want nothin' to do with it. I'm a trucker."

We decided that he would stick to the trucking and transporting aspect of it and Wayne and I would take care of the rest. Eventually each of our wives got shares in the company. After three or four years, after Phyllis and I divorced, Wayne wanted out. He got a new line of machinery for his shares.

Using my home in Merville as the office, we stocked and sold farm machinery, fencing and other related items which I'd truck over from the mainland. The business grew in leaps and bounds. There wasn't a lot of competition in the Valley then and that was one of the reasons it did so well. But above all, CJC succeeded because service is what sells. When you're right at home, customers always know where to find you. They can take advantage of buying twine after supper, or whenever is most convenient for them. I had a pretty good stock of parts and I can only remember about two or three instances when people didn't get fixed up when they came with something broken. The basic needs were always met.

I can remember this one Friday when Henning Offerson went to buy a part for his baler from a merchant in town. When he got there at 5 p.m., the store was just locking up. Henny was panicking a bit, because his hay was ready to bale and the store was closed Saturday and Sunday. He stopped at my place on the way home, I didn't have that particular part, but I had a baler ready to sell to him for $1200. I said, "I'll follow you home with it." He was out haying within twenty minutes, and by Sunday, when the rain began to fall, he had it all cut and safely packed into the barn.

For the most part with most of the stuff we bought we

either came out even or ahead, although there was the odd time when that wasn't the case. Once I purchased a neat old cash register for $200 and when I got it home I realized there was a two dollar bill jammed in the back. I paid a locksmith $40 to get the bill out and the machine never worked the same again. So I chucked it. Can't always win.

22

Army Surplus

I only sold CJC Equipment three years ago and soon after I did the same with the fencing business. It now belongs to Black Creek Farm and Feed. The machinery business I sold to the Andrew Brothers. They were all great ventures while they lasted, but as usual it was time to move on. In 1995, Stan Zwicker and I started up an army surplus store under the name B.S. Enterprise. We never even discussed the business really, it just sort of happened. I've known Stan for quite a few years, ever since our trucking days, and one day he stopped in on his way to Victoria. He was going down to bid on some things and I agreed to go along for the ride.

It's funny, when the bids are going up, some buyers can be as tight as a bull's ass at fly time. Not me. And on that particular day, Stan and I bought a lot. The next month we did the same. Before we knew it, we had quite a bit of stuff accumulated between the two of us. Over the years, we've bought just about everything. From backhoes and Cats, to ladies' stockings. Anything we think we can sell.

In the beginning we were selling the stuff out of an old house of mine, and out of Stan's place too. Now we supply the military store beside the Island Highway. It's quite the business even now and it keeps us out of trouble. The way I see it, especially with all of the used but new government office equipment we've bought and sold over the years, Stan and I are preventing landfill. Our newest venture this month is to buy a herd of buffalo. We'll see what happens with that.

23

Elk Hauls

O ver the years I've moved over one hundred and sixteen head of elk to different parts of Vancouver Island and not just for the money either. I do it for the challenge.

The first haul happened in 1983, when the local fish and game club hired me to transport elk from Gold River to Constitution Hill. The only place in North America, or maybe in the world, that has Roosevelt elk is the Olympic Peninsula and Vancouver Island. We were transporting them for hunting purposes, but we were also dispersing them for their own survival.

Transporting works best close to spring when the elk are hungry. We'd build a big corral with a gate, then strap a wire about half way down and bait it with alfalfa. Eventually an elk would come along, trip the wire, and a car battery hooked to an electric motor would slam the gate shut. Sometimes we'd only get a couple elk, other times we'd get lots. And sometimes, we could have the trap set there all winter and we wouldn't get a damn thing.

But usually, if we baited them good, we'd catch a load every week ranging from three to thirteen.

My very first time at it was a pretty comedic event. These

three biologists came up from Oregon to help out and I guess they had done it before. Those kind of people like to show authority. We met on the site at 7 a.m. to get the fourteen elk we'd trapped into the back of my truck. I got maybe one hundred yards from the corral when one of the guys jumped in front of me and said all in a panic, "Bud, don't go any closer. You'll spook the elk." I said, "You dumb asshole, how are we going to get them in the truck if we don't get near them?" The guy went silent and I was thinking to myself, "Too many people on the site at one time is what spooks the elk." I just knew I wanted to get the job done as quickly as possible. One of the other "authorities" insisted we have a meeting before we got started. "These animals are pretty wild," he said.

"The wilder they are, the easier they are to handle," was my reply as I jumped into my truck and got it lined up with the cattle loading squeeze. Time was wasting.

Determining the age of each elk was done by taking a tooth and a small blood sample. Then, their ears were tagged and radio collars were put on, to keep track of their location. I didn't want to fart around with this because I knew we had to be quick in order for me to catch the ferry to Louis Lake. I convinced the others that it would be quicker and easier if we crammed five or six of the animals into the corral first, where they'd be shoulder to shoulder and not be as able to move around. Then I'd crawl in on top of them and do the work twice as quick. The other guys were trying to blindfold the elk, along with all these other techniques, because they were scared of hurting the elk. "You're better off hurting them a little bit and getting the job over and done," I said. To me it just seemed crazy. I mean, these were educated people. They wanted to put collars on the yearling bulls and I was saying, "How do you make sense of that, when you know that its neck is still going to grow twice as big?" To that they suggested leaving lots of slack for growing room. I said, "Then the elk are in danger of getting caught up on a limb and choking to death." Then they wanted to tranquilize them. "I won't haul them if you tranquilize them," I said. "Just leave them be. If they need it, I'll do it."

I was feeling pretty impatient by the time they finally had the truck loaded and then there was an issue about transporting them through Campbell River in daylight. There were protesters who had already come up and burned the corral down a couple of times and the Game Club was worried about them being up in arms if they spotted me. They suggested I wait until evening to be as discreet as possible. "I ain't sleeping here," I said. "Get outta my way." And I jumped in my truck and headed down island. They were forced to follow me, but not too close behind. In the end, it all went pretty smoothly.

A couple of years later I did another transport, only this time I did it with my friend Bud Smith and some other guys. Doug Janz was a biologist out of Nanaimo and Craig Sawchuck, then the president of the local fish and game club, were both instrumental in this particular expedition. We were at Jordon River, preparing to unload seven elk. It was the middle of the night and to top it off, it had started to snow, so it was hard to know where we were.

When we finally came across the herd we had a hell of a time trying to load them up. Especially this one cow. She squealed like a baby crying while trying to bite us, but we finally managed to load her with the others and we were on our way. We took a load above Union Bay and two or three loads over to Port Alberni.

I had a hunch that the elk wouldn't be content long eating

from the bushes. They seemed more partial to the hay fields, and that's exactly what happened. The ones we dropped in Union Bay were soon busy browsing through people's lawns and hanging out on the ball diamond. Before we knew it, five or so got killed on the highway. I believe one swam across to Denman and is still there.

Not every mission was a success. There was this time I had to transport thirteen of them for MacMillan Bloedel and being superstitious like I am, I wasn't crazy about that number. I agreed to drive the load, but I assured the Port Alberni game warden that there would probably be a few fatalities. Sure enough, when we arrived two of the elk stepped out of the truck, laid down on the road and died right there. Stress I guess. They would have been fine if we'd gotten home quicker — I could have given them a shot of penicillin.

The last elk haul I did was in 1988. It's funny, for twenty eight years now I've entered my name in the island elk hunting draws and strangely I haven't won yet. Some years I've put in for eight different draws (deer, moose, elk, bear and wild turkey) and I still haven't won.

24

What is Love?

There's nothing I like better than hunting and sex. They're the two greatest things in life. In fact, I can't think of anything nicer without laughing. But to be honest, I don't really know what love is. I don't think I ever stopped working long enough to really think about it. I always thought that if they had money in their purse and could go buy whatever they needed, it made more sense than stroking their hair.

Phyllis left in 1977. We agreed to disagree. The divorce became final in 1980. She was getting over her sickness and the strain of the businesses got to be a bit much for her. The phone never seemed to stop ringing around our house, so I purchased another house down on Bates Beach, for a bit of an escape. I went away on a hunting trip with my son-in-law, Wayne and at about midnight, he dropped me off at the Bates Beach house. Phyllis was a very clean woman and wanted me to take my dirty clothes off on the porch. Well we got into a little argument over that and that lasted until about 2 or 3 a.m. I have to say, I wasn't too surprised when I received a letter from Phyllis' lawyer asking for a divorce.

I thought about what assets I had and I sure as hell wanted her to have half of them. I gave her a choice of the farm, or the house on Bates Beach, and money. She agreed to the latter and it was signed and sealed. The divorce dragged on for a few years and I kept working throughout. She sued me for $120,000 plus her lawyer fees, which I didn't have. It wasn't too long after that, she went after my pension. My lawyer shot into action and put a stop to that pretty quick. She got pretty hungry.

I don't think the divorce affected Dwight as much as Sharon, but I know it must have been hard on them. I didn't want to drag them into any of it. I've seen Phyllis very little since. Occasionally I'll bump into her over at Sharon's, maybe once every five or ten years. She doesn't want to talk to me too much and I don't really want to talk to her. I'd always said, when we were happily married, "If we ever split, the province is too damn small for the both of us." But we're still living in it. She did work damn hard over the years and she did a great job. Especially with the tack shop and the auction barn.

My second wife, Anne, and I had some good times over the nineteen years we spent together. Quality time. Lots of riding and hunting, and she traveled around with me on my buying sales. Anne seemed to enjoy working the stock, and she could handle any farm work that was thrown at her. We had quite a lot in common.

Aside from being a hard worker, Anne liked hunting and she was an excellent shot. We hunted all over B.C., especially in the Itcha Mountains. She once shot a caribou at four hundred and fifty yards, and it's rack-span measured eight inches wider than the last recorded in the Boone and Crocket Record Book.

Packing home eight hundred pounds of meat from a hunting expedition was not a bit uncommon. On one occasion the moose she shot was so big we had to pack his pieces out on three horses for three days. It was pretty cold out, but we were thankful that there weren't any flies — or wolverines, which I've heard are tough enough to back up a grizzly. When night fell the wolves came around and we had to shoot the few that tried to get their teeth into our fresh meat.

If there's one hunting trip that stands out most in my mind, it's the time Anne had to ride out with bread bags on her feet. It must have been in the early eighties. Anne, my buddy Roy, and I went on a trip around the middle of November on the second to last day of the season. When we set out riding in the early morning, it was colder than a clam's ass, let me tell you. At least minus ten on the old fahrenheit scale.

Riding through snow that was fourteen or fifteen inches high, we spotted a moose and shot it. While Roy dressed it, Anne suggested coffee and a fire. It wasn't yet nine o'clock so we decided to hunt a little longer. A mile later we came across a meadow scattered with another small herd. I jumped down from my horse, handed Anne the reins, and Roy and I each shot another moose. Anne's horse reared up when the shots went off. When she hit the snow her horse took off but she managed to hold on to mine. Roy's moose was wounded and had taken off into the bush, so he went off after it. Anne and I decided it was a good time for coffee.

We skinned the one moose, met up with Roy, and made a fire. Anne was sitting there warming her boots a bit too near the flames, when all of a sudden they just curled and melted right off her feet. Roy couldn't stop laughing but Anne got the last laugh later on when Roy was looking for the sandwiches. Anne said with a snicker, "Too bad, they went off with my saddle horse."

We had to go four miles back to the camp and Anne's boots were toast, so for a while she rode in her sock feet. But the next morning, while Roy and I were out hunting, she had found some breadbags, and she wrapped them over her socks. She rode all the way home with those bread bags on and she never complained. They were better than nothing, I guess.

25

No Regrets

THE GOLDEN YEARS

The golden years have come at last.
I cannot see, I cannot pee
I cannot chew, I cannot screw
My memory shrinks
My hearing stinks
No sense of smell
I look like hell
My body's drooping
Got trouble pooping
The golden years have come at last.
The golden years can kiss my ass.

There are things that have changed in my life over the years, but mostly things have remained the same. I still rise at the crack of dawn, because once my eyes are open I'm pretty energetic. My era is where the phrase "Up and at 'em" is from. If I have any regrets it's that I wasn't born sooner, before the government came along and screwed up this great country. I think it's gone to hell in a hand cart really. As long as the government can borrow money to keep floating and keep the do-gooders happy by buying their votes, well between that and the Indians, there's not a helluva lot left for the rest of us. At this point I just hope the Indians can settle their land claims so that there's enough land left for a white reserve.

And I don't regret not completing school. I don't feel it's ever held me back.

I've lost a lot of people that were close to me over the years. In 1978, my sister Mary passed away. I know that she drank gallons of coffee a day and one morning, after she got up to have a cup with her son, she went back to bed and died in her sleep. Two years later, Mother passed away. In 1982 my son Dwight died.

My old Airdale dog, Tobe, would have been one hundred and five in people years, when he passed away last year. He started digging a hole in front of the steps and I remember thinking to myself, "Shit, he's digging his own grave." So I buried him there. I guess that somewhere out there, there's a place for me too.

Bud Today

All photos in this section by Boomer Jerritt

At the old house

Me and my dog

Some of my trophies

At the old house

At the auction ring